Withdrawn

POWER PLANTS WITH
AIR-COOLED CONDENSING SYSTEMS

AIR-COOLED DIRECT CONDENSATION SYSTEM
146/160 MW UTRILLAS POWER PLANT, SPAIN

Reproduced with the Permission of GEA Gesellschaft fur
Luftkondensation m.b.H.

POWER PLANTS WITH
AIR-COOLED CONDENSING SYSTEMS

E. S. Miliaras

The MIT Press
Cambridge, Massachusetts, and London, England

Library of Congress Cataloging in Publication Data

Miliaras, E S
Power plants with air-cooled condensing systems.

Includes bibliographical references.
1. Electric power-plants. 2. Waste heat. I. Title
TK1041.M5   621.312'132   73-20101
ISBN 0-262-13093-9

PUBLISHER'S NOTE

The aim of this format is to close the time gap between the preparation of certain works and their publication in book form. A large number of significant though specialized manuscripts make the transition to formal publication either after a considerable delay or not at all. The time and expense of detailed text editing and composition in print may act to prevent publication or so to delay it that currency of content is affected.

The text of this book has been photographed directly from the author's typescript. It is edited to a satisfactory level of completeness and comprehensibility though not necessarily to the standard of consistency of minor editorial detail present in typeset books issued under our imprint.

The MIT Press

MONOGRAPHS IN MODERN ELECTRICAL TECHNOLOGY

Alexander Kusko, Series Editor

1. Solid-State DC Motor Drives by Alexander Kusko
2. Electric Discharge Lamps by John F. Waymouth
3. High-Voltage Measurement Techniques by Adolph J. Schwab
4. The Theory and Design of Cycloconverters by William McMurray
5. Computer-Aided Design of Electric Machinery by Cyril G. Veinott
6. Power-System Reliability Calculations by Roy Billinton, Robert J. Ringlee, and Allen J. Wood
7. Power Plants with Air-Cooled Condensing Systems by E. S. Miliaras
8. Magnetic and Electric Suspensions by Richard H. Frazier, Philip J. Gilinson Jr., and George A. Oberbeck

CONTENTS

CONTENTS

CONTENTS

CONTENTS

CONTENTS

FOREWORD

The electric power industry is strained by an increasing demand for electric energy in the face of restrictions on air pollution, thermal water pollution, radiation safety, and the difficulties of obtaining sites and rights-of-way for power plants and transmission lines. The air-condenser power plant, in which the waste heat of the exhaust steam is transferred directly to the air rather than to water, can reduce many of the industry's problems. The plant is most suitable for a mid-range generation of 1500 to 3500 hours per year, can be sited without consideration of cooling water, and can be placed near load centers without requiring additional transmission lines. The air-condenser concept is not new; it was not studied carefully in the U. S. because of apparently higher cost when ample cooling water was available for conventional systems. Several air-condenser plants have been built in Europe; more are planned as the concept is proven.

Mr. Miliaras treats all aspects of the air-condenser plant: history, heat exchangers, impact on plant and turbine design, economics, and system considerations. He relies on his own experience in the design of power plants, as well as on published designs and data in Europe and the U. S., to show the feasibility and desirability of air-condenser power plants. References in the U. S. and foreign literature will permit the reader to investigate the subject in considerable detail.

FOREWORD

This book will stimulate new thinking by the electric
power industry on an important power plant concept.
As compared with many ideas that have been proposed
to provide energy without degrading the environment,
but that require extensive research, the air-condenser
power plant can be built with today's technology for
tomorrow's electric power loads.

Alexander Kusko, Editor
Monographs in Modern
Electrical Technology

PREFACE

My interest in air-cooled power plants began nearly four
years ago with a brief study of their potential that also included a search of the meager--and at that time mostly for-
eign--literature.

The implications of adopting air cooling as the mode for waste
heat rejection from power plants, with the obvious reflection on
power generation costs and the more subtle effects on the
thermodynamic cycle and system planning, continue to occupy
my attention. This interest, I hope, is reflected in the present monograph.

While the information is presented in a form useful to the
power plant designer, the intention has not been to compile
a design manual on air-cooled power plants, since such an
effort is probably premature at this stage of their development,
but rather to present in a manner accessible to the engineer
with a not-too-distant acquaintance with thermodynamics and
heat transfer, the state of the art and potential improvements
and advances in the separate processes and components, and
their optimal thermodynamic and economic integration in a
power generation cycle.

This volume represents an individual effort, particularly in the
opinions expressed and the conclusions reached. At the same
time, of course, the literature and specifically the references
given at the end of each chapter have been drawn upon freely.
For permission to reproduce a number of figures and tables I
wish to thank the American Power Conference, the National
Committees of the Fifth, Sixth and Seventh World Power Con-
ferences, the Institution of Mechanical Engineers, the
American Society of Mechanical Engineers, CIGRE, the Heat
Exchange Institute, General Electric Company, Westinghouse
Electric Corporation, GEA Gesellschaft für Luft Kondensation

PREFACE

m.b.H., GKN Birwelco Ltd., Electrical World, Power and
the McGraw-Hill Book Company, Pergamon Press, Energie
und Technic, Verlag Chemie G.m.b.H., Electrizitatswirt-
schaft, and Combustion. The figures reproduced are
identified by a number indicating the original source
among the references at the end of each chapter.

It is a pleasure to express my thanks to Dr. Alexander Kusko,
Editor of the Series, "Monographs in Modern Electrical Technology," in which this volume is presented, for providing a
foreword.

My good friend and former associate, Mr. Robert Neery, has
been very kind in agreeing to review the typescript prior to
publication. My many bilingual friends, to whom I am in-
debted for assistance with foreign literature translations,
are probably much relieved with the publication of this volume, expecting an end to their enforced contributions, but
this will probably not be the case as air-cooled power plants
continue to be developed and find greater acceptance overseas.

My thanks also to Miss Norma Ricci, who has provided
valuable editorial assistance with the manuscript.

Chapter One

INTRODUCTION

The increasing size of power plants, competition for
adequate cooling water resources, thermal pollution,
scarcity of coastal and inland water sites, and the desirability for such sites near population centers have
all increased the demand for new concepts for the
rejection of waste heat from a power plant.

A comparatively new concept as applied to power
plants, the dry cooling system features finned-tube,
air-cooled heat exchangers rejecting heat to the air
either directly from the condensing steam or indirectly
through a spray condenser and hence via a circulating
water loop to an air-cooled, finned heat exchanger.
The heat exchangers can be incorporated in natural
draft hyperbolic cooling towers or mechanical draft
assemblies.

The direct system has been used on a limited scale
in Europe, while the indirect system has been pioneered
by Professor Heller in Hungary. Heat rejection to air
has been used extensively by the process industries for
refineries and process plants in arid areas. It has also
been used in cases where the large temperature difference between the condensing or circulating fluid in the
finned exchanger and the air made the system practical
over the wet, evaporative cooling system.

The lower-cost wet cooling system where heat is
rejected by water evaporation in an air stream, after
a slow start in the United States, has gained increasing favor in the last few years. However, its use may
have to be restricted because of high water usage and
the occasional formation of objectionable fog or mist
that creates siting problems.

The hardware cost of dry cooling systems utilizing finned exchangers is particularly high when compared to the conventional systems with surface condensers that are rejecting heat to circulating water.
While in the surface condenser heat is transferred
through the tube material and two high-value film coefficients, those for the condensing steam and the
circulating water, in the finned tube air-cooled heat
exchanger the heat rejection is controlled by the low-

value air side film coefficient, which is of the order of
a hundred less than the condensing steam or the waterside film coefficient. The resulting need to provide a
proportionally larger, fin-augmented heat exchanger surface with the dry cooling system and the requirement for
arranging and supporting such structures for improved
air access, has elevated the heat rejection end of the
power plant to a position of prominence, competing with
the turbogenerator and boiler as a major cost item.

As with a dry cooling system the power plant rejects heat to the ambient air, the dry bulb temperature is the appropriate design criterion. Since air is
usually warmer than cooling water supplies and as a
higher temperature difference is used between the heat
rejecting and heat receiving media to reduce heat exchanger size, turbine back pressures will be higher and
efficiencies lower for plants equipped with dry cooling
systems.

The novel emphasis placed on heat rejection and the
high investment and detrimental effect on efficiency of
the dry cooling system require that careful reconsidera-
tion be given to cycle analysis, optimization and equipment selection for a power plant.

The higher cost and lower efficiency of a plant with
a dry cooling system can be the only recourse for isolated systems in areas where there is no cooling or evaporative water available.

Occasionally, the disadvantages can be balanced
with savings in transmission line construction by locating near load centers or in fuel transportation by locating the plant near the fuel source.

Several aspects of power plant design have to be re-
evaluated on the basis of incorporating a dry cooling system in the cycle:

Methods for providing short duration overcapability
in the boiler and turbine or other means of compen-
sating for loss of capability in a power plant with
a dry cooling system at high seasonal ambient
temperatures.

The utilization of multipressure spray condensers, as they appear to offer large benefits with dry cooling systems.

Bottoming the power cycle with ammonia or the freons, that may offer advantages for dry cooling systems in cold climates.

Mid-range generation with steam power plants designed for peaking/cycling service, as they can utilize most of the advantages offered by the dry cooling systems, while reducing the disadvantages.

Dismal as the prospects may appear from the viewpoint of plant costs and efficiency, the dry cooling system offers the designer unusual opportunities for plant optimization, as it introduces another major variable, the heat rejection system, to be considered on a par with the boiler and the turbine in the cost and efficiency trade-offs. It also gives the utility system planner greater flexibility as it offers freedom from siting requirements near cooling water supplies and long transmission lines. It further offers an attractive new plant concept for serving intermediate and some peaking loads where the gas turbine has been the sole expensive choice.

The currently popular method-treating power plant optimization studies by completing a large number of heat balances on the computer and then searching, plotting and summarizing the results for trends-has been avoided here in favor of the analytical, approximate treatment of the effect of variables on plant performance. The latter provides insights and possibly avoids oversights all too common when only the final results of the computer's efforts are considered.

The present monograph is intended to provide a unified treatment of the subject by examining the effect of design parameters on dry cooling system and other major item hardware choices and by determining their effect on power plant cycle selection.

Chapter Two

THE DIRECT SYSTEM

2.1   Early Developments

Direct steam condensation inside finned air-cooled tubes for power generation was first introduced in 1939 at a coal mine power station in the Ruhr district of Germany. The installation, reportedly in operation until recently, is shown in the photograph, Figure 2.1. The largest installation to date, for the 160 MW Utrillas Power Station commissioned in 1970 in Spain, is shown in the photograph inside the front cover. Other installations, important in power generation, and their salient features are summarized in Appendix A. About one hundred direct steam condensing installations have been completed to date, the majority associated with smaller in-plant power generation units, with a total condensing duty of about ten million lbs. per hour.

Steam condensation by air is reported to have been actively considered in the early 1930's by several companies in England, Sweden and Germany with disappointing results except for GEA Luftkuhler Gesellschaft in Germany, which produced a design with oval, air-cooled finned tubes for a steam turbine at the above mentioned mine; the development work is reported to

Reproduced with the Permission of GEA Gesellschaft fur
FIGURE 2.1 Luftkondensation m.b.H.

The oldest air-cooled steam condenser in the world — in operation since 1939. Condensing capacity 12,500 lb/hr. Operating pressure 1.73 in. Hg.

have been carried out in Prandtl's laboratory in Göttingen.
The need for an air-cooled steam condenser was then
visualized not only for power generation in arid countries
but also for the purpose of conserving condensate in
steam-driven locomotives and avoiding frequent stops
for replenishing boiler makeup.

A parallel impetus for developing air-cooling equipment for refinery and industrial process applications arose
in the United States in the early 1940's, beginning with
natural gas processing and gas pipeline plants in the arid
southwest. From small applications for interstage cooling
of compressed gas, lubricating oil coolers, condensing
refrigerants and in fractionators, the present large module
U. S. designs for refinery and process applications
evolved with capabilities of several billions of BTU's
heat rejection per installation.

A development that somehow failed to generate any
subsequent results was the construction of several "power
trains" in the United States in the closing days of World
War II as sources of mobile power for war needs and subsequent reconstruction efforts. Each "power train" was
a modular power plant of 5 MW rating, mounted on railway
locomotives, and was equipped with direct steam air-
cooled finned condenser, the finned tubes and fans comprising the roof of two locomotives.

2.2 Steam Condensation Inside Tubes

The mechanism of vapor condensation inside a cooled
tube has received considerable study not only for steam
condensation but also for its applicability in reflux and
downflow condensers in the process industry. High rates
of heat transfer from the vapor to the tube wall are
realized, affected by vapor and liquid properties, vapor
velocity and tube geometry.

The following equation applies to saturated vapor
flowing at substantial velocities downward (in parallel
with the condensate) in a vertical tube. For upward
steam flow in a vertical tube (counterflow to the condensate) and for vapors substantially at rest, no simple
expressions for h_m can be derived and equations

yielding the heat transfer coefficient have to be solved
by numerical methods.

$$h_m = \frac{3k}{2}\sqrt[3]{\frac{ab}{H}} = \frac{3}{2}\sqrt[3]{\frac{\phi\gamma\lambda k^2}{3\mu H(t_f - t_s)} \cdot \frac{\rho_v v_v^2}{2}} \qquad (2.1)$$

Where:

h_m — Mean heat transfer coefficient across the liquid film, from the steam to the tube surface, averaged over the tube height H.

$\phi =$ — 4f, where f, the friction coefficient, is a function of Reynolds number. ϕ is proportional to $v_v^{-1/4}$ in turbulent flow.

γ — Specific weight of the liquid

λ — Latent heat of condensation

k — Thermal conductivity of the liquid

ρ_v — Vapor density

v_v — Vapor velocity

μ — Dynamic viscosity of the liquid

H — Tube height

t_f — Saturation temperature

t_s — Tube inner surface temperature

The original analysis of vapor condensation was
carried out by Nusselt. To illustrate the effect of vapor
velocity and pressure on the coefficient of heat transfer
to the tube wall, Jakob calculated the following values

from the modified Nusselt equation given above which are compared with values for steam at rest in Table 2.1.

For a vertical tube, with steam and condensate in parallel flow (downward),

a tube height   H = 3.28 ft. and $t_f - t_s$ = 18 °F

TABLE 2.1

Mean Value of Film Heat Transfer Coefficient "hm" for Steam, in BTU/hr ft^2-°F

Steam Pressure, psia	.15	14.7	73.5
h_m for Steam at rest	800	1000	1100
h_m for Steam at 500 ft/sec	1500	3600	6800

Several other rather complicated semi-analytical correlations, utilizing experimental data for heat transfer in condensation, have been subsequently developed, as indicated by the titles in the references, to compensate for deficiencies in Nusselt's analysis that account for measured values of the heat transfer coefficient about 20% higher than indicated by Equation 2.1.

In direct steam to air condensation applications, the situation is further complicated by the varying steam velocity with condensing duty and by the inclined orientation of the tubes that causes the condensate to flow preferentially in the lower part of the inclined tube rather than evenly cover the inside tube surface.

Some general statements nonetheless apply:

> The heat transfer coefficient for condensation is strongly affected by steam velocity, increasing with the .58th power of vapor velocity for parallel flow, as can be demonstrated from the equation, and it tends to decrease for the lower part of the tube.

> Counterflow condensation is less efficient than in parallel flow, with a considerably lower heat transfer coefficient.

2.3 Commercial Designs

The results of an experimental investigation of the
overall heat transfer coefficient, including tube and
air side film resistance when condensing steam in an
air-cooled finned heat exchanger and its variation with
air velocity and fan power, are shown in Figures 2.2a, 2.2b
and 2.2c. The assembly tested was a thin bundle of
staggered finned oval tubes 7.4" wide and 12.5 feet
long, inclined at an angle of 64 degrees. Saturated
steam was at .74 psia and 1.47 psia and the cooling
air was at about 70°F.
　　The tests were carried out during the development
of the German design widley utilized today
in standardized modules for direct steam condensation
for power generation and described in some detail here.
Several other manufacturers with extensive process
experience in the design and manufacture of finned vapor
condensing equipment are also offering standardized
designs of finned heat exchanger assemblies for direct
steam condensation.
　　The GEA modules referred to above are seen in the
photograph inside the front cover. There appears to be
little difference between these modules and the original
installation in Figure 2.1, but actually the modern standardized modules incorporate several advancements and
refinements in the state of the art, some proprietary
to the manufacturer. The finned tubes assembled in
panels are arranged in pairs forming A frames with the
fan at the base of each frame (module) providing the
forced air flow. The installation, comprising 40 modules
condenses 755,000 lbs. of steam per hour, at 3 inches
H_g rejecting $665 \cdot 10^6$ BTU/hr of heat to an airflow of
$1.4 \cdot 10^9$ cu. ft. per hour at 59°F and 3500 ft elevation.
The exhaust steam ducting and steam distributing headers
are easily identifiable.
　　From the upper headers steam flows downward and
condenses inside the finned tubes, giving up latent
heat to the cooling air. The condensate flows to the
lower, collecting headers that also pass the remaining
steam to the counterflow sections. The counterflow

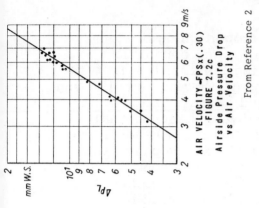

AIR VELOCITY—FPSx(.30)
FIGURE 2.2c
Airside Pressure Drop
vs Air Velocity

From Reference 2

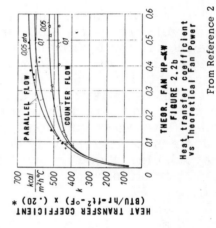

THEOR. FAN HP—KW
FIGURE 2.2b
Heat transfer coefficient
vs Theoretical Fan Power

From Reference 2

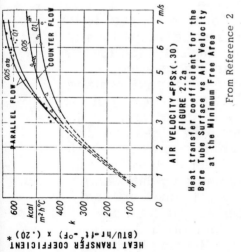

AIR VELOCITY—FPSx(.30)
FIGURE 2.2a
Heat transfer coefficient for the
Bare Tube Surface vs Air Velocity
at the Minimum Free Area

From Reference 2

*The multipliers have been added to convert the metric scales
of the figures to British System units.

sections have separate fans; proportioning of steam
condensing duty between parallel and counterflow
sections is performed by varying fan speed . From the
collecting headers the condensate flows by a collecting
pipe to the condensate tank and hence, via the condensate pumps, enters the feedwater heater circuit.
Non-condensables in the upper and lower headers are
piped to air ejectors, mechanical vacuum pumps that
compress and discharge them to the atmosphere. Vacuum
relief valves allow quick air inflow and draining of the
system to prevent freezing upon loss of load in cold
weather.

The smaller finned heat exchanger to the right, near
ground level, serves for auxiliary water cooling to the
lubricating oil and hydrogen coolers.

Figure 2.3 shows diagrammatic details of the finned
tube assembly for parallel and counterflow arrangements
and lists characteristics of each. In parallel flow, due
to the steam pressure drop in the tube in the direction
of steam flow, the saturation temperature of the condensate decreases progressively downward and subcooling
is unavoidable. Flooding and freezing problems and
oxygen absorption by the subcooled condensate are also
severe.

In counterflow the condensate gets heated as it flows
downward against the rising steam and nearly isothermal
condensation is realized. The heat transfer rate is lower
than in parallel flow and there is some steam loss and
overloading of the air extraction pumps at high loads.
Usually the two flow arrangements are combined in an
installation, with the final condensation taking place
in counterflow sections, eliminating subcooling of the
condensate and minimizing the other problems.

Steam maldistribution between the tubes and the
formation of inactive zones which occurs when the air
flows across successive tubes, favoring the first tube
in the air flow and increasing condensing duty there
while the surface of the outer tubes is underutilized,
can be ameliorated by varying the fin pitch on the
tubes, as shown in Figure 2.3. In another design,
the tube rows are drained to the collecting headers
via loop seals that compensate for the varying condensing rates and pressure drops in successive tube

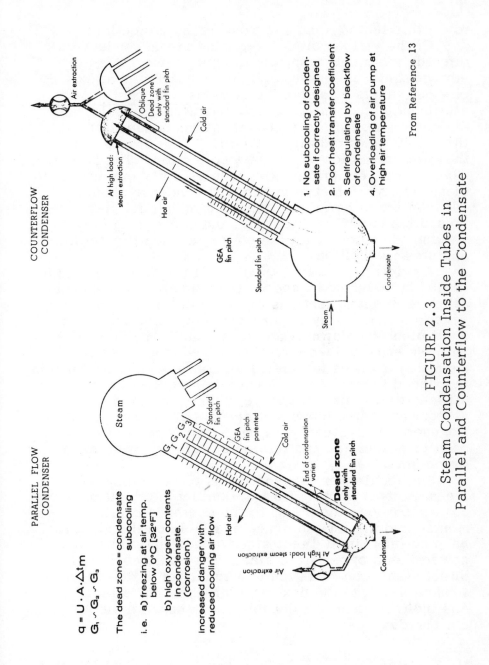

FIGURE 2.3
Steam Condensation Inside Tubes in
Parallel and Counterflow to the Condensate

rows; this feature also assuring freeze protection.

Of the various other air-cooled steam condensers offered, the giant mushroom in appearance, Rosette* arrangement, is shown in Figure 2.4. The heat exchanger panels of carbon steel tubes with continuous aluminum plate fins, two per fan-powered module, are assembled around a center-core steam header from which the steam flows radially outwards. The lower part of the core serves for condensate collection. The arrangement shown, comprising six modules, is about 40 feet in diameter and 36 feet high and will condense about 100,000 lbs of steam per hour at 8 inches H_g steam pressure and 80°F ambient air, the condensing duty varying of course, with fan power, ambient temperature and steam condition.

Air flow to the air-cooled condenser is usually controlled by varying fan speed; at reduced plant load and at low ambients, sections of the installation comprising several modules may be shut down and isolated. For short plant shutdowns vacuum is usually maintained, with isolating valves sometimes used to prevent steam migrating from the turbine seals condensing as ice inside the finned tubes. Ample air removal equipment is utilized to evacuate the large volume system after longer shutdowns, when tubes and headers are filled with atmospheric air. Steam purging the air from the system prior to evacuation and turbine startup has been found expedient.

The requirement for short runs of large diameter steam exhaust ducting from the turbine, to limit pressure drop in carrying the steam to the overhead cooling elements, and the limited space available on the plant roof have so far limited direct steam condensation by air to power plants of moderate size. The applicability of direct steam condensation to large units is discussed in more detail in the next chapter where the direct and indirect steam condensation systems are also compared.

*U.S. Patent 3,630,273. "Rosette" is a General Electric Co. Trademark.

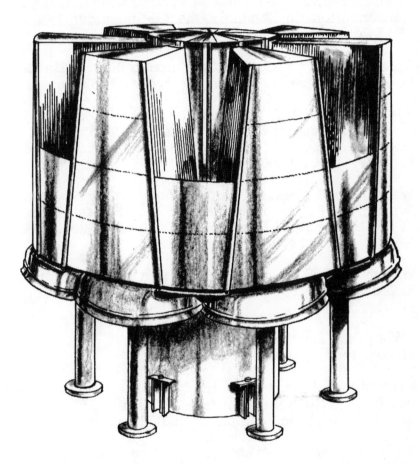

FIGURE 2.4
Direct Steam Condensation
Air-Cooled Heat Exchanger Arrangement

REFERENCES

1. Bowman, R.A., Air-Cooled Steam Condensers; Trans-
    actions of the A.S.M.E., Vol. 67, Nov. 1945.

2. Schulenberg, F., Air Instead of Water for Cooling and
    Condensation (in German); Chemie-Ingenieur-Technik
    25. Jahrg. 1953/Nr.10.

3. Oplatka, G., Air-Cooled Condensing Plants; The Brown
    Boveri Review, Vol. 49, No. 7/8 Aug. 1962.

4. Van der Walt, N.T., Discussion to "A Closed Circuit
    Cooling System for Steam Generating Plant" by L.J.
    Cheshire and J.H. Daltry; The South African Mechanical
    Engineer, Feb. 1960.

5. Wood, B., Discussion to the "Rugeley Dry Cooling Tower
    System" by P.J. Christopher and V.T. Forster; Proceedings
    of the Initiation of Mechanical Engineers, Vol. 184, Pt 1,
    No. 11, 1969-1970.

6. Weyrmuller, G.H., Modular Design of Air Exchangers
    Saves Refinery 20% in capital costs; Chemical Processing
    January 1970.

7. Jakob, M., Heat Transfer, Volume I, Chapter 30, John
    Wiley, Inc., 1949.

8. Carpenter, F.G. and Colburn, A.P., The Effect of Vapor
    Velocity on Condensation Inside Tubes; 1952 Proceedings
    of the General Discussion of Heat Transfer.

9. Rohsenow, W.M., Webber, J.H. and Ling, A.T., Effect
    of Vapor Velocity on Laminar and Turbulent-Film Con-
    densation, Transactions of the A.S.M.E., Vol. 78,
    1956.

10. Dukler, A.E., Dynamics of Vertical Falling Films Systems,
    Chemical Engineering Progress, Oct. 1959.

11. Dehne, M.F., Air Cooled Overhead Condensers; Chemical
    Engineering Progress, Vol. 65, No. 7, July 1969.

REFERENCES (Cont'd.)

12.   Bell, K.J., Amplification, Qualification, and Correction
      (Discussion of preceeding reference);Chemical Engineering
      Progress, Vol. 66, No. 4 April, 1970.

13.   Gessellschaft für Luftkondensation m.b.h. (G E A)
      Bochum,W. Germany; Company Publications.

14.   Von Cleve, H. H., Westre, W. J. and Parce, J. Y.,
      Economics and Operating Experience with Air-Cooled
      Condensers; Paper presented at the American Power
      Conference, Chicago, Ill. April 1971.

15.   Simpson, N., Air-Cooled Condenser Fits Steam Plant
      to Arid Site; Electrical World, June 8, 1970.

16.   Von March, F., Rziha, H. and Kelp, F., Planning and
      Erection of the Air-cooled 160 MW Steam Power Station
      in Utrillas (in German); Brennst-Warme-Kraft, Volume
      22, No. 7, July 1970.

17.   Schoonman, W., Air-Cooled Steam Condensers for Power
      Plants; Paper presented to the (Australian) Society of
      Mechanical Engineers Symposium on Decentralization
      of Energy Production in Relation to Available Water
      Resources, Sydney, Australia, Sept. 1970.

18.   Rosette Air-Cooled Steam Condensers; Heat Transfer
      Products Dept., General Electric Company.

Chapter Three

THE INDIRECT (HELLER) SYSTEM

3.1 Introduction of the Indirect Steam Condensation System by Prof. Heller

The indirect system was proposed by Professor Heller in 1950, at the Fourth World Power Conference in London. The date marks the beginning of a lively interest in large-scale heat rejection to air from power generation sources. The system is outlined in the flow diagram, Figure 3.1. Elements of the system are (1) the spray condenser, a large vacuum-tight shell equipped with water spraying nozzles, (2) the water to air finned tubular heat exchanger located in the airflow generated by a natural draft tower or by fans, and (3) the circulating water pumps that pump the cooling water to the tubular exchangers at above atmospheric pressure through the circulating water piping, and the head recovery turbines that recover part of the pumping energy before the circulating water is sprayed at subatmospheric pressure in the spray condenser. (The last item is usually replaced by throttling valves in smaller installations.)

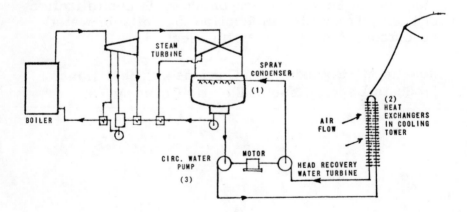

FIGURE 3.1
Schematic Flow Diagram of a Power Plant
with an Indirect Air-Condensation System

The turbine exhaust steam enters the spray condenser and condenses by direct contact with the sprayed circulating water from the tubular water-to-air heat exchangers. From the condenser hotwell the mixed condensate is divided in the ratios of between 1 to 20 to 1 to 50 into the condensate stream, that often after further deaerations and treatment enters the feedwater heater train to the boiler, and the larger circulating water stream that is cooled in the tubular water-to-air heat exchangers and returned to be sprayed again in the condenser. The ratio of the condensate to the circulating water stream determines the cooling range in the water-to-air heat exchangers.

The Heller system is perhaps the most important original concept in the last quarter century in power generation, a field that has not been characterized for encouraging innovations.

It utilizes the spray condenser, an inexpensive piece of equipment from the early power plant days, to replace the expensively tubed surface condenser. Purity of the mixed condensate-circulating water from air contaminants and oxygen -- impossible to achieve if the spray condenser was to be utilized with a conventional wet cooling tower -- is accomplished by enclosing the mixed condensate-circulating water in tubular finned heat exchangers.

As the efficient evaporative heat transfer process of the wet cooling system is replaced with inefficient heat transfer to air through a high-resistance air boundary layer, Dr. Forgo, a Heller associate, developed the tubular water to air heat exchanger with slotted-rib fins that break up the boundary layer and improve heat transfer rates. Of course, the higher Dry Bulb temperature is the design parameter for the dry cooling system instead of the Wet Bulb temperature that the circulating water approaches in the familiar wet cooling tower. A small advantage is gained as the 5°F minimum terminal difference required between circulating water and condensing steam temperatures in a surface condenser is eliminated in the spray condenser.

Heat rejection to air makes power generation possible far from sources of cooling water and in arid areas where evaporative cooling water for use with wet cooling towers is scarce. It allows locating the power plant near the load center or at the source of cheap fuel. Professor Heller foresaw two advantages for the indirect system over the direct steam-to-air condensation system that also achieves the same goals:

> As the water pressure inside the tubular heat exchangers is above atmospheric, contamination of the condensate by air inleakage is avoided and tube leaks can easily be detected by the resulting visible water spray.

> The need for large-diameter ducts to carry the large-volume steam flow from the turbine exhaust to the cooling elements is eliminated.

The first advantage is not substantiated on the basis of experience with direct steam-to-air condensation systems, where tube faults and air in-leakage have not been a problem. But the elimination of the large-diameter steam exhaust ducting by condensing in the spray condenser under the turbine, and the efficient transport of the rejected heat by the circulating water to the cooling elements at some distance from the turbine, made possible the removal of the finned tubular heat exchangers from the confines of the power plant roof and introduced air cooling to large units.

3.2 Large Indirect Condensation Systems and Their
     Operation

Following development work on a 1.2 MW pilot plant, the first practical application of the Heller system was in a 16 MW plant at the Danube Steel Works in Hungary, commissioned in March 1961. A photograph of the plant with a schematic diagram of the tower arrangement is shown in Figure 3.2. In December 1961 the well-known

From Reference 3

FIGURE 3.2a
The Danube Steel Works Plant in Hungary (16MW-1961).
First Indirect System Air-Cooled Plant.

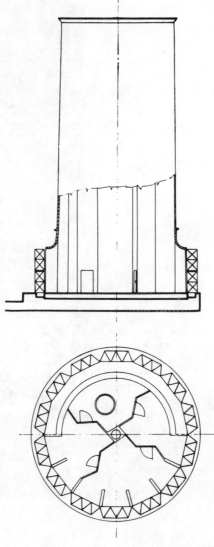

From Reference 3

FIGURE 3.2b
Arrangement of the Dry Cooling Tower.
Danube Steel Works Plant.

120 MW unit with the first large Heller system of steam condensation at the Rugeley Station of the CEGB in England was commissioned, and in April 1967 a 150 MW unit was commissioned at the Preussag Plant at Ibbenbüren in Germany. Recently (in 1971) the Grootvlei 200 MW unit in South Africa has been completed, while the two 200 MW units at Gyongyos in Hungary and the three 220 MW units at Razdan in the U.S.S.R. are in various stages of commissioning or approaching completion. Salient features of these plants are included in Appendix A.

All the Heller system plants completed so far utilize natural draft cooling towers, ideally suited for the large air flows required. In the United States, where the traditionally low fuel costs have favored mechanical draft, several indirect system designs by manufacturers have been proposed with the air-flow to be provided by very large diameter fans.

At both the Rugeley and the Ibbenbüren installations, the finned tubular exchanger assemblies are grouped in pairs to form A-frames and arranged on their sides along the periphery of the tower in a zigzag pattern with the long axes vertical, as shown in the Ibbenbüren tower photograph, Figure 3.3.

The cooling system is sectioned into four quadrants, the heat exchangers in each quadrant comprising an independent circuit. Two half-sized circulating water pumps operating in parallel, each on a common shaft with a head recovery turbine, supply the quadrants.

The head recovery turbines are usually omitted in smaller installations, the excess pressure head, above that required for the sprays, between the pressurized cooling elements and the condenser vacuum taken across a throttling valve, with a resulting small inefficiency. At Ibbenburen, as in the smaller Danube Steel Works installation of Figure 3.2, the interior of the tower is also sectioned, to the height of the cooling elements, to prevent crossflow at high winds.

Water storage tanks of adequate capacity to contain the total volume of the water in the system are located

From Reference 8

FIGURE 3.3
Indirect System Air-Cooled
Heat Exchanger Modules

at the base of the tower or indoors. Quick draining of
the cooling elements is accomplished by isolating the
particular elements and allowing air through the relief
valves located at the top of the assemblies, as seen in
the photograph, Figure 3.3.

To prevent freezing of the circulating water in the
cooling elements during cold weather, the water temperature is continuously monitored and maintained above a
certain level, usually 41 to 43°F. Control is accomplished by isolating and draining sections at low loads,
shutting down one of the two circulating water pumps,
and at Ibbenbüren by controlling the airflow through
the tower by adjusting louvers located at the outside
face of the cooling elements.

As the steam turbine exhaust area is usually sized
small to better perform at the higher back pressures
encountered with dry cooling systems, choking of the
last turbine stage will prevent utilization of lower
vacuums to improve performance, and reduction of
vacuum below a certain point is unwarranted. In installations utilizing forced draft, shutting off fans below
a certain ambient at a given load, rather than lower
vacuum, will improve overall performance by reducing
the substantial auxiliary power requirements of such
installations.

Complicated piping arrangements provide for isolating,
draining and filling the cooling elements, the latter
function performed during periods of cold weather with
condensate from the storage tank preheated by precir-
culation in the condenser. A schematic of the piping
arrangement for the Ibbenburen plant is shown in Figure
3.4. To simplify the hydraulic circuitry, multi-ported
"sector valves" especially developed for that purpose
were used at Rugeley.

In this context a proposed piping arrangement is of
interest, the top of the cooling elements maintained at
subatmospheric pressure by connection through a vent
pipe to the spray condenser. To drain, a three-way
valve isolates the section from the condenser vacuum
and allows venting, the incoming air dropping the water
level in the elements by the effective height of the condenser vacuum, about 25 feet. To refill the section,

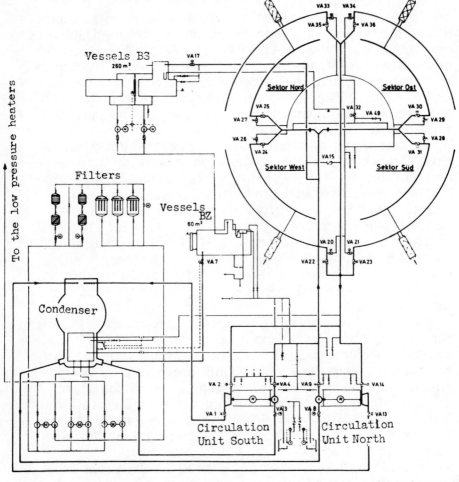

FIGURE 3.4
Condensate-Circulating Water Flow Diagram
for the 150 MW Plant at Ibbenbüren

the three-way valve closes off atmospheric access and reconnects to the condenser, the vacuum there quickly raising the water level in the elements. Of course, one of the original advantages claimed by the Heller system is lost, i.e., maintaining the cooling elements under pressure, but as already mentioned, air in-leakage in a vacuum system presents no great problem. The arrangement considerably simplifies the piping circuitry, and as the circulating pump head requirements are lowered by not pressurizing the elements, the head recovery turbines can be eliminated.

Both Rugeley and Ibbenbüren utilize all-aluminum heat exchangers, while at the recently completed Grootvlei plant carbon steel fins on carbon steel tubes, with the assemblies hot dipped galvanized, have been used. Also the arrangement of the assemblies differs, as the A-frames are located inside the tower with the long axes horizontal. To prevent internal corrosion of the carbon steel tubes, nitrogen in a closed system is used to displace the water when sections are drained; nitrogen replaces the water and is conserved in the water storage tanks as the operation is reversed when the sections are refilled.

Potential problem sources in the Heller system plants -- flooding of the condenser if a circulating pump stops and water hammer to which the system is susceptible -- have also been ingeniously provided for by the designer.

At start-up, to initiate draft, opposing quadrants are usually brought up together. Vacuum is reacned quickly, in about ten minutes, as for a surface condenser. To simplify operation, automation and control function centralization has been extensively used, along with the European practice of subdividing the power plant control system into independent functional subgroups, each with a mimic diagram for operator information and guidance. This contrasts with the American practice of a single computer monitoring and centrally annunciating all functions. The following excerpt from the next to the last reference given amply illustrates the scope and simplicity of operation at the Ibbenbüren station.

"This air condensation plant is provided with up-to-
date automatic controls in order to facilitate running the
station. In fact, all maneuvers concerning running up,
shutting down or performance control, consisting each
of several part operations, are carried out simply by
pushing a button. The automatic control equipment first
checks whether the conditions of carrying out the command are fulfilled, ensures the correct sequence of operations and indicates the possible disturbances. It is
thus possible to start up or shut down automatically any
of the two circulating pumps and water turbines normally
operating in parallel, and the valves involved in filling
or emptying the quadrants of the water system divided
into four parts also operate automatically as required.
There is also an automatic sequence control program for
the case of a valve not functioning properly because of
some disturbance or defect and thus preventing the ful-
fillment of the command given.

"Air Condensation plants have to operate under widely
varying weather and load conditions requiring adequate
performance control from time to time. Due to the nature
of power station operation, however, manual operation
control proved to be adequate, especially as facilitated
by the automatic sequence controller described. Thus
air flow through the cooling tower is controlled according to changing conditions by means of the louvres provided, the air flow being reduced in cold weather, and
at low loads one of the two circulating pumps normally
running in parallel is shut down. As far as airflow control is concerned it was found convenient to divide the
louvres into several parts, the upper third being operated
manually and the rest by suitable actuators from the control room. Thus, operation control consists of a combination of maneuvers in function of the load and the weather conditions. The decisions to be taken by the operating staff are facilitated by the combined instrument
shown. This has two measuring systems, one of which
indicates the actual electrical output of the turboset,
the other the prevailing temperature of the atmospheric
air. The needles of the two instruments cross as shown
and the position of their intersection over the indicating
field indicates the optimal plant conditions in the actual

situation. Since sudden load or weather changes are
unusual, there is normally ample time for carrying out
the operation by the automatic control equipment."

The instrument referred to above for operator guidance
at Ibbenburen is shown in Figure 3.5.

Following some early difficulties not uncommon in
plants utilizing new equipment, the operation of the
Rugeley and Ibbenbüren plants has been eminently successful from a performance and reliability standpoint.

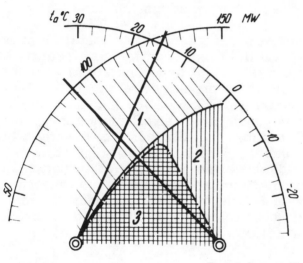

From Reference 14

FIGURE 3.5
Instrument for the Control of the Air Cooling System at
Ibbenbüren

1. Two-pump operation, louvres open

2. Upper tier of louvres closed

3. Single-pump operation

Early external corrosion in the finned heat exchangers
at Rugeley, because of the combination of humidity and sea-
borne chlorides , has been extensively researched and docu-
mented. A specially developed coating is used on the
new assemblies as the original heat exchangers are being
replaced. Condensate subcooling was also eliminated
by rearranging the spray condenser internals.

At Ibbenburen, malfunction of the vent valves prevented complete venting during the first winter of operation,
the resulting impediment to circulation causing freeze
damage requiring plugging of several tubes and replacement of one assembly. All the vent valves were replaced
with a new design by the Hungarian suppliers and steam
traced.

Following these early problems, operation at both stations has been reliable and continuous, meeting all design
goals. The Rugeley plant has consistently carried over-
loads when grid requirements were severe and has recently
been used successfully in cyclic, two-shift operation.
Ibbenburen, because of better efficiency compared to the
older plant available that utilizes wet towers, is usually
operating at baseload with capability maintained to 95 °F.

The individual elements of the Heller System, the
associated operational requirements and the implications
to utility system planning of the adoption of steam condensation via heat rejection to air, are discussed in the
chapters that follow.

3.3 The Direct and Indirect Steam Condensation Systems
      Compared

The merits of the direct versus the indirect system of
condensation have been the subject of some conjecture,
the consensus appearing to foresee the adoption of the
direct system for the smaller generating units, up to
perhaps 200 MW rating, while preempting the larger
unit ratings to Heller System application, including of
course all nuclear plants.

A superficial comparison of the characteristics of the
two systems does not yield their specific merits or
explain the attractiveness of the Heller System for the

larger units. Heat transfer coefficient that determines
heat exchanger surface, which is the dominant cost item
with both systems, is controlled by the air side film coefficient common to both systems for the same tube and
fin configuration and air velocity, irrespective of steam
or water flowing in the tubes. The waterside film coefficient for the Heller System also has about the same
value as the condensing steam coefficient for the direct
system, discussed in the last chapter.

The waterside film coefficient for the Heller System
heat exchangers, as calculated from the familiar Dittus-
Boelter equation* for the range of parameters of interest,
is shown in Figure 3.6. The improvement of the waterside film coefficient at the higher circulating water temperatures is primarily due to the lower viscosity of the
warmer water. Higher coefficient values are usually
realized than indicated by the equation if there is internal tube roughness or corrugations, such as sometimes
result from the application of the fins to the tube. In
this regard, internally grooved tubes are currently suggested as a means for improving heat transfer and reducing tube surface in conventional surface condensers, and
the method may have some marginal applicability to the
Heller System heat exchangers. Heat transfer rates,
measured with a Forgo fin heat exchanger bundle, are
shown in Figure 5.3, they may be compared with the
rates for a direct steam condensation heat exchanger,
Figure 2.2.

Superficially, then, on the basis of heat transfer coefficient consideration, it may appear that the air heat
exchangers' capital cost will be about the same for the
two systems, with the indirect system further burdened
by the additional cost of the spray condenser, the circulating water pumps and head recovery turbines, as
against the higher cost of the large-diameter steam exhaust and distribution ducts of the direct system and
the latter's evacuation and steam distribution problems.
But the circulating water piping and auxiliary power requirements for the Indirect System's water pumps are less
sensitive to condensing steam pressure than the Direct
System's exhaust duct size and pressure drop which are
functions of steam specific volume, increasing inversely

* McAdams, Heat Transmission, 3d. Edition; p. 219.

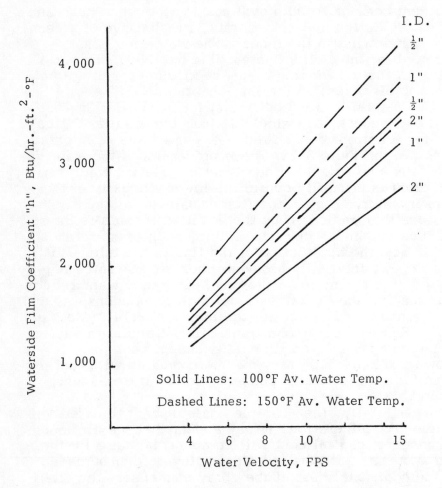

I.D.

½"

1"

½"

2"

1"

2"

Solid Lines: 100°F Av. Water Temp.

Dashed Lines: 150°F Av. Water Temp.

Waterside Film Coefficient "h", Btu/hr.-ft.²-°F

Water Velocity, FPS

FIGURE 3.6
Waterside Heat Transfer Coefficient
for Indirect System Heat Exchangers

with decreasing condensing pressure. With the low condensing pressures associated with the European designs,
it appears that the two systems are on a par for overall
economy for plants at about 200 MW, with the Direct
System more economical for smaller plants and the Indirect System having the advantage for larger plants. At
the higher back pressures that will be associated with
U.S. designs, the Direct System will probably prove
economical for plants of considerably larger rating.

Cascading the indirect system's spray condensers
also offers benefits with large units. Steam-to-air or
water-to-air heat exchanger surface for given heat rejection load is about inversely proportional to the Initial
Terminal Difference (ITD) between the media, plant efficiency also varies inversely with ITD, which determines
turbine back pressure. For a given plant, ITD is determined by utility system considerations, such as fuel
economics, load factor, cooling system and other plant
equipment costs, and should be about the same for the
Direct or Indirect Systems. But with the Indirect System,
for a given air-cooled heat exchanger surface, ITD can
be "effectively" reduced and efficiency improved by reducing the average condensing steam temperature through
condensation in series connected spray condensers. (Figure 4.7 and Section 7.5); conversely for the same plant
efficiency as the single spray condenser installation air-
cooled heat exchanger surface can be reduced.

There is a valid thermodynamic reason for the improvement with the cascading arrangement of the spray condensers, as discussed in detail in a later chapter. Multi-
pressure spray condensers, the mixed circulating water-
condensate from the first condenser sprayed into a second
at a higher steam pressure, and so on, are practical with
large rating units utilizing four and six flow low-pressure
turbine sections.

Another reason for the advantage of the Heller System
with large units is the suitability of the associated heat
exchanger design to natural draft installations. The
Forgo low air-side pressure drop aluminum slotted-rib
heat exchanger, identified with the Heller System, has
contributed to improved overall performance.

The following exerpt from the last reference listed also compares the indirect and direct condensation systems and their performance.

"Thermo-Dynamic Differences Between The Two Systems: Fig. 1 (Figure 3.7, next page) gives a brief schematic comparison between the direct and indirect systems. The water temperature in the fin tubes of the indirect system decreases, whereas the condensation temperature in the direct system remains nearly constant, which gives a better logarithmic mean temperature difference (LMTD) and thus a smaller cooling surface required for the same heat quantity to be removed."

"The performance characteristics of both systems, namely the turbine back pressure being a function of the air temperature, are nearly equal. The influence of the pressure drop in the exhaust pipe of the direct system, however, results in lower steam pressure at high heat load and high air temperatures, and vice versa, Fig. 2." (Figure 3.8, next page).

"The individual advantages of the two systems are of an economic nature, and depend on the size of the plant and the distance between turbine and condensing plant. In the range of 200 and 300 MW turbine output, (about 200 to 300 gigacalories per hour, or 800 to 1200 million Btu per hour) a careful economic comparison should be made for both systems. Larger plants have to be located at a larger distance away from the turbine house, so that the indirect system has economic advantages, because the total cross section of the connecting pipes necessary for the direct system is 2.5 to 3 times larger and the pressure drop in the exhaust steam pipe has an influence on the total size of direct system plants whereas in indirect system plants the pressure drop influences only the auxiliary power. For smaller plants, the simple direct air-cooled system which needs no accessories such as large storage tanks, pumps, jet condenser, water turbine, etc. is more economic. Nevertheless, direct air-cooled condensers have been designed in detail for plants up to 300 MW using natural draft towers. The special advantage of natural draft direct air-cooled condensing plants is that no auxiliary power is needed at all, thus reducing the annual costs for the total plant."

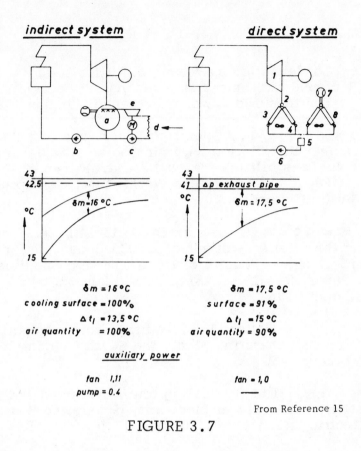

FIGURE 3.7

From Reference 15

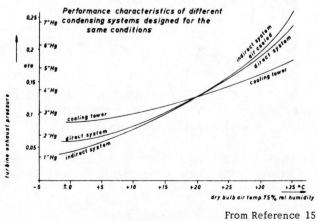

From Reference 15

FIGURE 3.8

REFERENCES

1. Heller, L., Condensation by Means of Air for Steam
    Turbines Equipped with Injection Condensers; Section E3,
    Paper No. 7 Fourth World Power Conference, London 1950.

2. Heller, L. and Forgo, L., Operating Experience of a Power
    Plant Condenser Installation with Air Cooled Circulating
    Water and the Possibilities for Further Development;
    Section G/3 Paper 113 G3/2 Fifth World Power Conference,
    Vienna 1956 (in German).

3. Heller, L. and Forgo, L., Recent Operational Experiences
    Concerning the "Heller System" of Air Condensation for
    Power Plants. Latest Results of Developments; Subdivis-
    ion $III3_2$, Paper 154 $III3_2/8$, Sixth World Power Confer-
    ence, Melbourne 1962.

4. Cheshire, L.J. and Daltry, J.H., A closed circuit cooling
    system for steam generating plant; The South African En-
    gineer, February 1960.

5. Christopher, P.J., The dry cooling tower system at the Rugeley
    power station of the Central Electricity Generating Board;
    English Electric Journal, Vol. 20, No. 1, Jan/Feb 1965.

6. Christopher, P.J. and Forster, V.T., Rugeley Dry Cooling
    Tower System; Proceedings of the Institution of Mechan-
    ical Engineers, Vol., 184, Pt. 1 No. 11, 1969/70.

7. Reti, G.R., Dry Cooling Towers; Proceedings of the American
    Power Conference, Vol. XXV, 1963.

8. Scherf, O., The Air Cooled Condensing Plant for the 150
    MW unit of the Preussag Power Plant at Ibbenburen;
    Energie und Technik, Juli 1969 (in German).

9. Heller, L., New Power Station System for Unit Capacities
    in the 1000 MW Order; Acta Technica Hungarica, 1965.

REFERENCES (Cont'd.)

10.   v. Cleve, H.H., Westre, W.J. and Parce, J.Y., Economics and Operating Experience with Air-Cooled Condensers; Paper presented at the American Power Conference, Chicago, Ill., April 1971.

11.   Heeren H. and Holly W., Air Cooling for Condensation and Exhaust Heat Rejection in Large Generating Stations; Paper presented at the American Power Conference, Chicago, Ill., April 1970.

12.   Smith, E.C. and Larinoff, M.W., Power Plant Siting, Performance and Economics with Dry Cooling Tower Systems; Proceedings of the American Power Conference, Vol. 32, 1970.

13.   Rossie, J.P. and Cecil, E.A., Research on Dry-Type Cooling Towers for Thermal Electric Generation; Environmental Protection Agency Water Quality Office, Report 16130EES11/70, November 1970.

14.   Forgo, L., The Heller System of Condensation by Means of Air for Power Stations. Operating Experiences and Development Objectives; VII World Power Conference, Moscow 1968.

15.   v. Cleve, H.H., Dry Condensing and Cooling Plants: Their Design, Characteristics and Economics; The South African Engineer, April 1969.

Chapter Four

EXTENDED SURFACE HEAT EXCHANGERS

4.1 Hardware Considerations: Common Types of Finned
     Surface, Construction, Materials, Codes, Exper-
     ience

Extended surface heat exchangers are utilized for heat
rejection in air-cooled condensing system applications
in order to improve air side heat transfer; the air side
heat transfer film coefficient is about two orders smaller than the inside tube film coefficient for the water or
the steam in the air-cooled heat exchanger.
     In this chapter hardware description and considera-
tions will be taken up as well as the relationship between air-cooled condenser surface area and plant efficiency for a given heat rejection load, while fan power
and natural draft tower requirements will be considered
in a following chapter. Heat transfer coefficients,
fin effectiveness, and fouling factors are briefly discussed in the next chapter.
     Of the various forms of extended surface available-
spines, studs, longitudinal fins, etc.-only uniform
thickness continuous helical and plate fins are generally considered for air-cooled condensing system application.
     Several common finned tube arrangements are shown
in Figure 4.1, while the cross sections in Figure 4.2
show details of fin-to-tube attachment. In general, two
fin types predominate: The endless helical fin, tension-
wound or extruded, and the plate fin. Details of attachment of tension-wound fins are shown in Figure 4.2a
through 4.2e, while the extruded fin, Figure 4.2f, results from slipping a tube of soft material, the muff, over
the inner tube and generating the fin by rolling the muff.
The inner tube surface is completely protected and the
rolling results in a tight interference fit between the tubes
and low contact resistance. Plate-type fins can be flat,
rippled or with turbulators, punched-in-protrusions that
break the boundary layer, creating turbulence and improving heat transfer. Plate fins can be individual to each
tube, Figure 4.1b, or common to several tubes penetrat-
ing the same fin plate, as in Figure 4.1a. Fins and spac-

From Reference 6, Chapter 3

FIGURE 4.1a
Finned Tubes of the Air-Cooled
Indirect Condensation System Heller

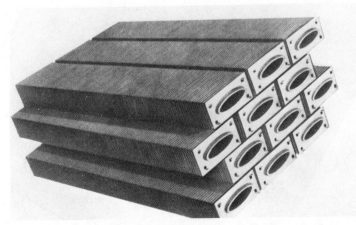

From Reference 3, Chapter 2

FIGURE 4.1b
Finned Tubes of the Direct Condensation
Air-Cooled System by GEA

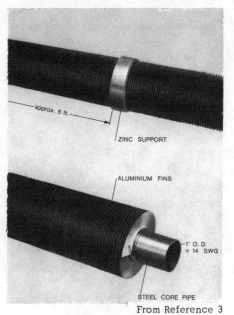

From Reference 3

FIGURE 4.1c
Finned Tube Details

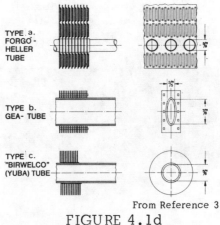
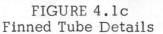

From Reference 3

FIGURE 4.1d
Finned Tubes Common with Air-Cooled
Heat Exchangers

ers are slipped and assembled over the tubes and the tubes are drawn internally, expanding outwards and creating a tight fit.

Usual tube sizes vary around one inch in diameter with 8 to 10 fins per inch and with the tubes usually staggered in the direction of the airflow. The requirements of air-cooled heat exchanger design for power plant condenser applications, i.e., large air flow with low pressure drop, the high inside tube film coefficient and the small change in the liquid temperature, favor shallow tube bundles of from two to six tubes deep in the direction of the air flow.

As header construction is expensive long tube lengths are favored, limited only by pressure drop and transportation requirements.

Two well-known fin types are the Forgo fin, Figure 4.1a, (aluminum slotted-rib plate fins on aluminum tubes that were used at Rugeley) and the GEA design, Figure 4.1b, widely used for direct steam condensation (the fin tubes pickled and then hot dip galvanized for good thermal conductivity and a flexible and durable bond between tube and fin). Increased surface for the same cross section as a cylindrical tube, some lowering of the resistance to the air flow, and less liklihood of damage in the case of accidental freezing are claimed for the latter, justifying the increased cost of manufacturing and handling the elliptical tubes.

The finned tubes are attached to distributing and collecting headers by a variety of methods that satisfy the low pressure requirements of dry cooling systems. Several designs are shown in Figure   4.3.

The fabricated carbon steel box header is quite common. Extruded aluminum headers are likely in the future. Tube ends, free of fins, are rolled or expanded into the headers. Access to the interior is provided through handholes for cleaning, scale removal, or occasional plugging of a leaky tube. Side channels, the length of the tubes, connect the headers to form a rigid structure, prevent the tube bundles from sagging when spanning wide bays, and protect the fins during shipment, at the same time allowing the relative movement of the headers due to thermal expansion during service.

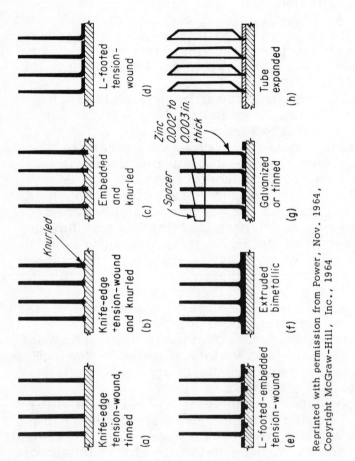

FIGURE 4.2
Fin Attachment Details

Knife-edge tension-wound, tinned
(a)

Knife-edge tension-wound and knurled
(b)

Embedded and knurled
(c)

L-footed tension-wound
(d)

L-footed-embedded tension-wound
(e)

Extruded bimetallic
(f)

Galvanized or tinned
(g)

Tube expanded
(h)

Knurled

Spacer

Zinc 0002 to 0003 in. thick

Reprinted with permission from Power, Nov. 1964,
Copyright McGraw-Hill, Inc., 1964

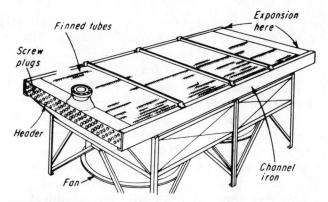

TUBE BUNDLE is made up of finned tubes roller expanded into headers at both ends. Screw plugs provide access to individual tubes

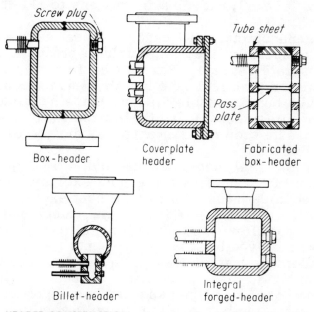

HEADER CONSTRUCTION varies primarily with fluid design pressure. Units are built for pressures reaching 12,000 psi and over

Reprinted with permission from Power, Nov. 1964,
Copyright McGraw-Hill, Inc., 1964

FIGURE 4.3
Details of Finned Tube Assembly
and Header Construction

Headers are usually designed to Section VIII of the ASME Code for Unfired Pressure Vessels. The code is not specific for box-type designs and the manufacturers often rely on proprietary information developed from testing. ASME and AIChE are currently active in preparing field test codes for air-cooled equipment, while ARI (Air Conditioning and Refrigeration Institute) and ASHRAE have developed standards for testing air-cooled finned exchangers and for certifying manufacturers in cooling and refrigeration applications (ARI Standard 410 in conjunction with ASHRAE Standard 33-64).

For process and refinery air-cooled heat exchangers, American Petroleum Institute Standard 661 applies (API Standard 661, Air-Cooled Heat Exchangers for General Refinery Service), with tube dimensions to API Standard 640.

The manufacture of finned tube heat exchangers lends itself to assembly line, large volume production, with commensurate savings. From a manufacturing standpoint the maximum practical tube length is most economical. Large tube bundle panels also minimize handling and labor for site assembly, while transportation and tube-side pressure drop requirements may dictate limits as to the size.

Carbon steel tubes and fins and various grades of aluminum tubes and fins are almost exclusively the materials considered for finned tube heat exchangers for dry cooling systems. With aluminum tubes and fins, the differential expansion and bimetallic contact problems are resolved but the requirements for control of the condensate may cause aluminum to go into solution. This problem is discussed further in Chapter 8. With aluminum tubes and fins, Rugeley has experienced severe corrosion between the fin collar and the tube surface; fin brittleness and wastage have also been experienced, all blamed on the corrosive chloride atmosphere and the high humidity of the site. Several other installations with aluminum fins have experienced no problems.

Bonding of aluminum to carbon steel is to be avoided. At Ibbenbüren, rubber hoses connect the carbon steel condensate piping to the all-aluminum heat exchanger tubes; but bimetallic combinations of aluminum fins tightly fitted on carbon steel tubes, the tubes internally expanded after insertion to the fin plates, are often encountered in the process industry and apparently present no problems other than the possibility of increased contact resistance due to differential expansion.

Tube surface protection and the integrity of the tight fit and good thermal contact between fin and tube are areas requiring major attention. Hot dip galvanizing of carbon steel tubes and fins has proven very effective if its integrity can be safeguarded during handling and erection. With aluminum fins, epoxy and other coatings of the assembled finned tubes are recommended for severe environments, such as Rugeley.

Fouling of the interfin surfaces has been experienced in process applications at several southern installations in the United States, from cottonwood and similar fouling material from floral sources in the spring; while simple accumulation of dust on the fins does not effect performance, as the resulting increase in fin efficiency tends to balance the insulating effect of the dust. Simple hose washing has proven effective for removing the dust. Internal tube fouling, common in the process applications, does not often arise with the condensate quality water required in power plant practice.

Tube air inleakage and the resultant deterioration of vacuum in direct system applications and deaeration problems with both direct and indirect systems because of the large number of tubes required in the heat exchangers--much discussed in the recent past as an obstacle to air cooling for power plants--can be dismissed on the basis of experience with large air-cooled process applications in the United States and the many foreign air-cooled power plant installations.

Large finned tube heat exchanger installations for process heat rejection are quite common in the petrochemical, process and refinery field; where their development to replace wet cooling towers received the strongest impetus. The experience gained, although at a higher Δt, is directly applicable to power plants; installations rejecting the equivalent thermal load of a 500 MW plant, but at higher temperature difference than preferred for power plant practice, are not uncommon in refinery service. The attitude towards life ex-

pectancy of the equipment varies: chemical industry
equipment costs are often written off in five years because of frequent process changes, while a required
thirty-year life is common practice in the utility industry.

4.2 The Selection of a Finned Surface Design by the
    Manufacturer and the Economic Evaluation by the
    Consulting Engineer

Heat exchanger details, such as ratio of finned to bare
tube surface, fin spacing, tube diameter, etc., are not
determined by the power plant designer who will usually ask for quotations from manufacturers, specifying
the heat load to be rejected at the design ambient and
the capitalized cost* of 1 kW in auxiliary power requirements for driving fans and pumps.
    On the basis of usually proprietary data, the manu-
facturer selects the heat exchanger design to be offered,
i.e., surface configuration and the resulting pressure
drops, by trading off first cost (exchanger surface and
required fans, pumps and motors) vs. operating cost
(fan and pump power in the case of a mechanical draft
installation). The selection is shown by the solid
lines in Figure 4.4. If the air-cooled heat exchangers
are to be utilized in an indirect system natural draft installation, the only operating cost variable to be optimized by the heat exchanger manufacturer will be circulating pump power, which is of relatively secondary importance, and the optimization will be mainly between
two first costs. A smaller, more compact, less costly
surface configuration will result in higher air side pres-

*Capitalized auxiliary power cost is the sum of the incre-
mental first cost for increasing the plant capability by 1
kW to provide the auxiliary power without changing the
net plant capability, and the capitalized operating cost
that depends on fuel cost, heat rate, operating hours over
the life of the plant, operating and maintenance labor, and
the fixed charge rate.

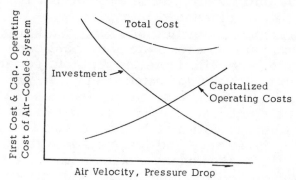

FIGURE 4.4
Air-Cooled Heat Exchanger Design Selection
by the Manufacturer

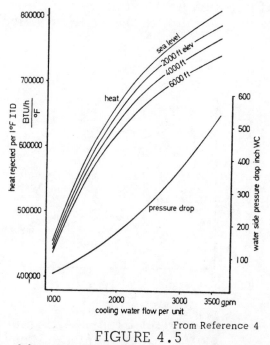

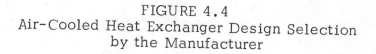

From Reference 4

FIGURE 4.5
Data Furnished by the Manufacturer for an Indirect System
Air-Cooled Heat Exchanger Module (Figure 6.1)

sure drop and a larger and costlier natural draft tower,
usually to be supplied by another manufacturer. That
the above optimization procedures do not include all
factors will be discussed later.

The manufacturer furnishes information in the form
shown in Figure 4.5 for a standard heat exchanger and
fan module. He will quote the required surface (and fan
power) to reject the specified load at the design ambient
at each of several turbine back pressures. Means for
calculating part load performance, together with finned
surface details, will also be furnished.

With the surface and pressure drop relationship
fixed by the heat exchanger manufacturer, the power
plant designer will select the optimum size air-cooled
system by trading off turbine backpressure, plant efficiency, and reduced plant capability at higher design
back pressures against air-cooled system size and cost,
as shown in Figure 4.6.

We will return now to the question of heat exchanger
design optimization by the manufacturer, Figure 4.4, and
the trade-off between heat exchanger surface (first cost)
and pressure drop, which in the case of a mechanical

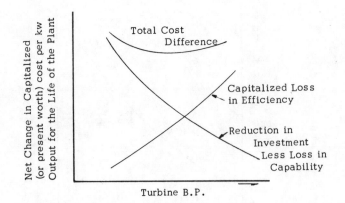

FIGURE 4.6
Air-Cooled Heat Exchanger Evaluation
by the Power Plant Designer

draft system will determine fan power (operating cost) while in a natural draft system will determine natural draft tower size (another first cost); information relating to the cost of pressure drop having been supplied to the manufacturer by the power plant design engineer.

Heat exchangers have been optimized in the past on the basis of several criteria: weight, volume, minimum frontal area, and also minimum investment and operating cost, as in Figure 4.4. Considerable literature on the subject exists and conflicting statements abound on the merits of various fin designs and the improvements in heat transfer by devices that reduce the boundary layer, such as turbulators, rippled fins, etc. The subject is further discussed in the next chapter.

We will assume that the manufacturer has available "in house" valid surface vs. pressure drop relationships on which to base his selection from the various heat exchanger surface designs he is considering for a specific job. That selection method will be perfectly adequate for the process industry heat exchangers, where the only constraint is that of minimum investment and operating costs in rejecting a given amount of heat with specified design temperatures. But for a power plant installation, where reducing the cold water temperature (or "tower approach" equation 4.7) will result in a lower turbine backpressure, increased plant output and efficiency, and lower heat rejection for the same electrical output, the overall system effect should be taken into account in the heat exchanger optimization by the manufacturer, not at a later stage in the optimization by the power plant designer. This can be done by furnishing the manufacturer a separate credit per degree of reduced cold water temperature, or by adjusting the cost of the auxiliary power given the manufacturer for optimizing a mechanical draft installation by the effect the lower approach will have on plant output and efficiency.

Analytical and computer studies optimizing heat exchanger surface design on this basis indicate that maximum airflow and minimum air side pressure drop should be favored, limited only by considerations of flow stability at the extremely low pressure drops.

4.3 Performance of Finned Surface Heat Rejection Systems

Heat transfer coefficient calculations, helpful in evaluating the adequacy of the various designs are discussed in Chapter 5, while pertinent statistics from several installations are included in Appendix A. Here the general relationship between heat exchanger size and backpressure will be considered.

Heat transfer between air and water (or condensing steam) in crossflow is mathematically described by the following equations and the diagram in Figure 4.7:

$$Q = F.U.A. \ LMTD \qquad (4.1)$$
for the heat transfer surface

$$= W_a(.24)(t_{a2} - t_{a1}) \qquad (4.2)$$
for air

$$= W_w(1)(t_{w1} - t_{w2}) \qquad (4.3)$$
for water

$$= w_s \cdot h_{fg} \qquad (4.4)$$
for condensing steam

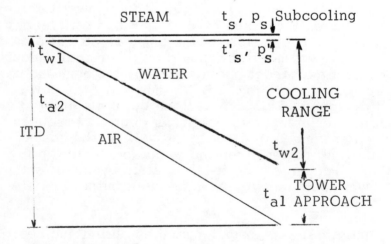

FIGURE 4.7
Temperatures of Air and Water (or steam)
in an Air-Cooled Heat Exchanger

Where:

Q	The amount of heat to be rejected by the air-cooled condensing system in btu/hr.
F	A geometric factor, near unity, to account for the crossflow heat transfer and often incorporated in "U" below.
U	The overall heat transfer coefficient in btu/hr-ft^2-°F, based usually on outside bare tube surface, or on the total finned surface. Sometimes also the heat exchanger frontal area is used as the basis.
A	The area in ft^2 on which "U", above, is based.
LMTD	The log mean temperature difference, in °F, between air and water or steam.
ITD	The difference between the initial temperatures of air and water or steam, $t_{w1} - t_{a1}$ or $t_s' - t_{a1}$, sometimes also referred to as "air-cooled condenser approach" to turbine exhaust steam temperature. The relationship between LMTD and ITD is indicated in Figure 4.8.
W_a	The air flow rate is lbs/hr.
W_w	The water flow rate is lbs/hr.
W_s	The condensing steam rate in lbs/hr, the major portion of the turbine exhaust steam flow as determined by the quality of the exhaust steam.
.24 & 1	Respectively the specific heats of air and water in btu/lb-°F.
t_{a1}, t_{a2}, t_{w1}, t_{w2}	Respectively the entering and exit temperatures of air and water; ($t_{w1} - t_{w2}$ is referred to as the "cooling range" of an indirect air-cooling system).
h_{fg}	The change in enthalpy of the condensing steam which varies slightly with turbine backpressure.

t_s, p_s Steam temperature and pressure at the turbine exhaust flange in °F and inches of Hg.

t'_s, p'_s Steam temperature and pressure prevailing in the direct system heat exchangers or the indirect system's jet condenser.

A difference exists, usually of the order of 1°F in adequate designs, between saturated steam temperature t_s corresponding to backpressure p_s at the turbine exhaust flange and average temperature t'_s at pressure p'_s in the heat exchangers of a direct system, or the exit temperature of the mixed condensate and circulating water of an indirect system's jet condenser. In its absence, turbine backpressure would have been improved by $p_s - p'_s$, with increased plant output and efficiency; conversely, for the same plant performance a larger ITD (and LMTD) would have been available, reducing cooling system size and cost. The effect of a fixed amount of subcooling becomes less pronounced at the higher ITD range.

The relationship between outside bare tube surface A and ITD for an air-cooled system rejecting 10^9 btu/hr has been plotted in Figure 4.8 for various air and water flows for the convenient value for the heat transfer coefficient of 100 btu/hr-ft^2-°F for the outside bare tube surface.

The bottom curve shows the ideal case--large volume flows of air and water (or isothermal steam condensation without pressure drops in the exhaust duct and heat exchanger); because of the large flows, the fluids do not change temperature as they exchange heat and the Initial Terminal Difference is present throughout the process (ITD=LMTD).

From Figure 4.8 it can be seen that increasing ITD sharply reduces the required surface to reject a given heat load; surface is actually reduced in inverse proportion to increasing LMTD (from Equation 4.1). As ITD is increased it approximates LMTD, as also is the case in general for the low air and water rises in existing installations. The following inverse proportion between

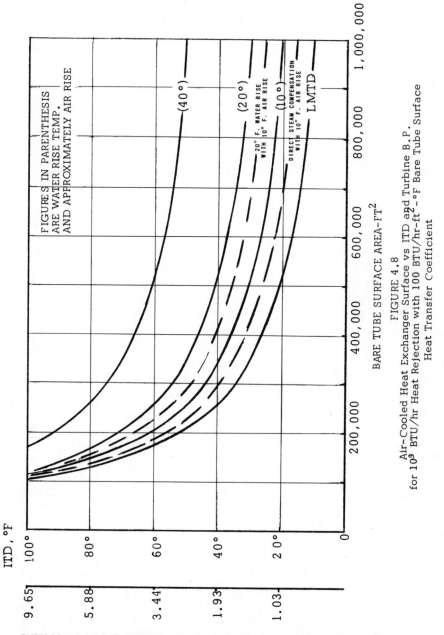

FIGURE 4.8

Air-Cooled Heat Exchanger Surface vs ITD and Turbine B.P.
for 10^9 BTU/hr Heat Rejection with 100 BTU/hr-ft^2-°F Bare Tube Surface
Heat Transfer Coefficient

Extended Surface Heat Exchangers

the frontal surface of the natural draft tower heat exchangers and ITD has been suggested:

$$Q = \frac{A_f \cdot ITD}{\frac{1}{K_1 V_a \rho} + \frac{PH}{K_2 n V_w} + \frac{1}{K_3 n U}} \qquad (4.5)$$

where,

A_f	The frontal area of the heat exchanger.
V_a	Air velocity.
ρ	Air density.
P	Number of water passes.
H	Height of the heat exchanger.
n	Number of tube rows.
V_w	Water velocity.
U	Overall crossflow heat transfer coefficient.
K_1, K_2, K_3	Constants.

It appears then that a convenient way to reduce heat exchanger size and cost, particularly for plants with a low fuel cost or a low capacity factor, i.e., cases where plant efficiency is not very important, will be by selecting higher backpressures and ITD. It also should be borne in mind that, aside from reduced plant capability at the higher design ITD, the heat to be rejected by the air-cooled heat exchanger per unit plant output also increases, as shown in Appendix B for various power plant cycles.

Other useful expressions relating heat rejection load to ITD and "tower approach" are:

$$Q = \frac{ITD(e^z - 1)}{\frac{e^z}{.24 W_a} - \frac{1}{(1) W_w}} \qquad (4.6)$$

$$Q = \frac{(t_{w2} - t_{a1})(e^z - 1)}{e^z \left(\frac{1}{.24 W_a} - \frac{1}{(1) W_w} \right)} \qquad (4.7)$$

Where, the exponent

$$"Z" = F_g UA \left(\frac{1}{.24 W_a} - \frac{1}{(1)W_w} \right)$$

and $t_{w2} - t_{a1}$ = "tower approach", the approach of the cold water temperature leaving the dry system's heat exchangers to the ambient dry bulb temperature, analogous to the approach of the cold water leaving an evaporative cooling system to the ambient wet bulb.

The relationship

ITD = Tower Approach + Cooling Range + Subcooling

can be surmised from Figure 4.7.

The following equations have been suggested to relate ITD to heat rejection load for existing installations:

$$ITD = AQ^{.75} \qquad \text{(4.8) for natural draft installations}$$

$$ITD = BQ^{.91} \qquad \text{(4.9) for mechanical draft installations}$$

Where A and B are constants depending on the heat exchanger design and the draft generating equipment.

From the above equations it can be seen that ITD may be assumed proportional to heat rejection load over a fairly wide range--more so for a mechanical draft installation. Rising ambient temperature reduces air density and in direct proportion the air mass flow rate delivered by the fans or generated by a natural draft tower (equation 6.1a), for constant pressure drop through the heat exchangers. But such pressure drop is reduced at higher ambients--nearly in proportion with density for constant mass--while the rising air viscosity affects both pressure drop and the heat transfer coefficient, resulting in roughly unchanging ITD with changing ambient over a wide range.

To summarize the effect of changing load and ambient on design ITD, we quote from the third reference

a foreign manufacturer of air-cooled condensing systems:

   "If the heat load is constant, the total temperature difference will remain the same and the
   saturated steam temperature will vary in the
   same measure as the outside air temperature.

   "During partial-load operation, the total temperature difference, with other conditions
   being equal, will vary roughly in direct proportion to the heat load."

The rules quoted were stated for a natural draft
indirect system with constant circulating water flow.
They generally also apply to mechanical draft installations (with no change in fan speed) as well as to
direct condensing air-cooled systems. More specific
performance information for the air-cooled mechanical
draft indirect condensation system with changing heat
rejection load and ambient is supplied by a domestic
manufacturer in Figures 4.9a and 4.9b below:

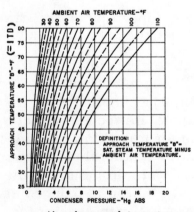

Air cooler approach temperature as a
function of condenser pressure and ambient air
temperature.

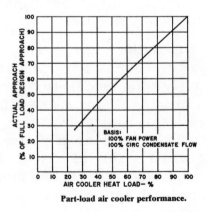

Part-load air cooler performance.

From Reference 12, Chapter 3

Figure 4.9a   Figure 4.9b

Incidentally the unchanged effectiveness* of the air-cooled condensing system contrasts with once-through cooling, where the minor seasonal variations of the cooling body of water are to some extent reduced by the improved effectiveness of the surface condenser (reduced condenser terminal difference with increasing inlet water temperature). Also wet tower performance improves at higher wet bulbs (reduced approach), with the result that the lesser variations of wet bulb temperature, as compared to dry bulb changes that affect air-cooled power plant performance, are considerably cushioned by the improved effectiveness of the wet tower-surface condenser system.

---

*Heat transfer system "effectiveness" defined as the ratio of the actual heat transfer rate to the ideal maximum possible rate in counterflow with infinite heat transfer surface.

REFERENCES

1. Howarth, F., Extended Surfaces: Their Use, Methods of Manufacture and Properties; The Chemical Engineer (Great Britain), June 1962.

2. Winters, G., Air-Cooled Heat Exchangers; The Chemical Engineer (Great Brittain), June 1962.

3. Heeren, H. and Holly, W., Air Cooling for Condensation and Exhaust Heat Rejection in Large Generating Stations, Proceedings of the American Power Conference, Vol. XXXII, 1970.

4. v. Cleve, H. H., Westre, W. J. and Parce, J. Y., Economics and Operating Experience with Air Cooled Condensers; Paper presented at the American Power Conference, Chicago 1971.

5. Kidd, A.C., Discussion on the "Rugeley Dry Cooling Tower System" by Christopher, P.J. and Forster, V.T.; Proceedings of the Institution of Mechanical Engineers, Vol. 184, Pt. 1, No. 11, 1969-1970.

6. Elonka, S., Air-Cooled Heat Exchangers; Power, November 1964.

7. Nakayama, E. U., Find the Best Air Fin Cooler Design; Petroleum Refiner, April, 1959.

8. Rabb, A., Are Dry Cooling Towers Economical?; Hydrocarbon Processing, Volume 47, No. 2, February, 1968.

9. Gerstmann, J., Reducing Costs of Dry Cooling Towers for Electric Power Plants; Power Systems Engineering Group Report No. 13, School of Engineering, Massachusetts Institute of Technology.

10. Andeen, B. R. and Glicksman, L. R., Computer Optimization of Dry Tower Heat Exchangers; ASME Paper 72-WA/Pwr-8.

Chapter Five

HEAT TRANSFER AND PRESSURE DROP WITH EXTENDED
SURFACES

5.1 Conductance Components and Fin Geometry

The overall heat transfer coefficient that determines the
size, i.e., heat transfer surface, and investment for dry
cooling systems is dominated by the film coefficient on
the air side, in both the direct and the indirect systems.
The condensing steam coefficient is in the range of 1000
Btu/hr ft^2 °F, while the water film coefficient for the indirect system, dependent on water velocity, is also in
the same range.
    The air side coefficient, dependent of course on air
velocity, is of the order of 10 Btu/hr ft^2 °F for bare tubes,
and this hundredfold disparity dictates the use of large
fins on the air side with the ratio of fin-to-tube surface
in the range of between 15 and 25 to 1. To complete the
comparison, representative values for the remaining heat
transfer coefficients in the heat path between the two
fluids are as follows: Waterside tube fouling coefficient
1,000, tube material 10,000, fin to tube contact coefficient 2,000, all in Btu/hr ft^2 °F.
    The overall heat transfer coefficient "U" for a thin wall-
ed finned tube, based on inside tube surface, is given by
the equation:

$$\frac{1}{U_i} = \frac{1}{h_i} + \frac{1}{h_{if}} + \frac{t}{k} + \frac{1}{h_c'} + \frac{1}{h_o'} \qquad (5.1)$$

Where
    h_i the inside film coefficient, steam to the tube
                   wall or water to the tube wall.
    h_{if} inside tube fouling coefficient.
    t tube thickness, ft.
    k tube material thermal conductivity
                   BTU/hr ft^2 – °F – ft
    h_c' contact coefficient, fin to tube
    h_o' equivalent air side film coefficient. Based on
                   inside tube surface, the equivalent air side
                   film coefficient reflects the improvement due
                   to the addition of the fins.

Heat Transfer and Pressure Drop With Extended Surfaces

The above coefficients are expressed, unless otherwise stated, in BTU/hr ft^2 °F. Those customarily expressed or measured on the basis of outside tube diameter, i.e. h'_c and h_o, are referred to inside tube surface by multiplying by the ratio of the diameters. The inside tube film coefficient "h_i", for the direct and indirect systems has been considered in the respective chapters; here primarily air-side considerations will be discussed, together with fouling and contact coefficients; finally, proposed schemes for improving airside heat transfer will be discussed.

Materials commonly used in air-cooled heat exchangers are carbon steel (k = 26) and aluminum (k = 118); by comparison copper's thermal conductivity is 220. In Table 5.I, from the third reference by Forgo, the specific weight (column 1) conductivity (column 2) and their ratios (column 3) are given in metric units for copper, aluminum and carbon steel and compared to that of aluminum taken as base (column 4). The prevailing average price for the materials in Hungary at the time is given in column 5 and the specific weight to conductivity ratio vs. price is compared in column 6 with the aluminum again as base.

TABLE 5-I

Comparison of Specific Weight, Conductivity, Price and their Ratios for Copper, Aluminum and Steel

Material	Spezifisches Gewicht kg/m^3	Wärmeleit-fähigkeit λ kcal/m °C h	γ/λ kg h° C/m^2 kcal	$\gamma/\lambda \big/ (\gamma/\lambda)_{Al}$	Material-preis Ft/kg	Material-preis bezogen auf Aluminium
	1	2	3	4	5	6
Kupfer	8900	320	27,8	1,80	29,0	2,3
Aluminium, 99,5%	2700	176	15,4.	1,00	22,6	1,0
Unlegierter Stahl	7850	43	182,0	11,80	11,3	5,9

From Reference 3

Heat Transfer and Pressure Drop With Extended Surfaces

As pointed out in the reference, varying manufacturing costs
for the materials complicate the comparison; also the material conductivity is but a minor part of the overall conductance for the finned tubes.

The most economical fin cross section is that of a tall
fin with parabolic sides that results in constant heat flux
(Schmidt 1926), with the tall thin fins commonly in use a
manufacturing compromise.

To describe the improvement in airside heat transfer because of the additional (but lower temperature) surface in
contact with the air when adding fins to the outside surface
of a tube, the concept of "fin efficiency" is employed. It
is defined as the ratio of the actual heat transferred from
the fin to the heat which would have been transferred if the
entire fin were at its root temperature or, otherwise stated,
if the fin material had infinite conductivity.

$$\emptyset = \frac{\sqrt{h}\ (t-ta)\ dA}{\sqrt{h}\ (tr-ta)\ dA} \qquad (5.2)$$

Where
   $\emptyset$ the fin efficiency
   dA   a differential element of the fin surface
   h   the airside film heat transfer coefficient
   t   the temperature of the fin surface element
   ta   the ambient air temperature
   tr   the temperature of the fin root.

For the uniform thickness integral annular fin--among other
shapes--Gardner developed an analytical expression of the
fin efficiency, shown plotted in Figure 5.1.*

*The upper curve in Figure 5.1 applies to straight fins, and
approximately to low fins where the ratio of fin height to
tube radius is small, and is sometimes used to calculate
fin efficiencies in such cases as it can be expressed in
the simple mathematical relationship:

$$\emptyset = \frac{\tanh\ (L)\sqrt{h/ky}}{L\sqrt{h/ky}} \qquad \emptyset_s = 1 - \frac{Af}{At}\ (1 - \emptyset)$$

Where "L" is the fin height and "y" is the fin halfthickness,
both in feet. Based on fin efficiency, a surface efficiency
"$\emptyset_s$" is also sometimes defined as the weighted efficiency
of the fins and the tube surface not covered by fins.

Heat Transfer and Pressure Drop With Extended Surfaces

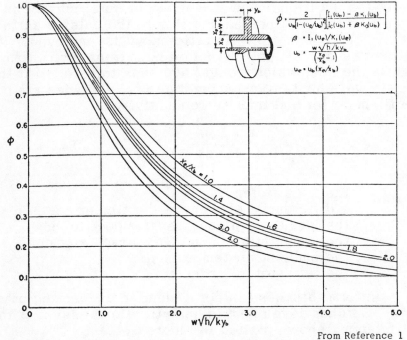

From Reference 1

$I_0(u)$: Modified Bessel Function of the first kind, zero order
$I_1(u)$:   "   "   "   "   "   ", first order
$K_0(u)$:   "   "   "   "   " second kind, zero order
$K_1(u)$:   "   "   "   "   "   ", first order
k: Thermal Conductivity of fin material

FIGURE 5.1
Efficiency of Constant Thickness Annular Fins

Heat Transfer and Pressure Drop With Extended Surfaces

On the basis of the fin efficiency, an equivalent airside fin coefficient is derived in the following form:

$$h'_o = h \cdot \frac{A_f}{A_i} \; (\frac{A_t}{A_f} - 1 + \phi) \qquad (5.3)$$

Where the subscripts "i", "f" and "t" refer to the inside tube surface, fin surface, and total fin and outside tube surface, respectively.

5.2 Airside Heat Transfer and Pressure Drop

The mathematical relation given for fin efficiency and the subsequent equation for the equivalent airside fin coefficient leave the impression that all ambiguity is removed when considering the merits of various finned surfaces; in actuality, the physical picture introduces several complica-tions.

Two phenomena present in the flow of liquid around a tube are of particular importance: The transition of the boundary layer attached to the surface from laminar to turbulent, with a sharp increase in heat transfer, and the subsequent downstream flow separation and generation of a wake behind the tube that increases pressure drop but contributes little to heat transfer.

Elliptical cross-section, plate-finned tubes, with the long axis parallel to the air flow, are claimed to be super-ior to round-finned tubes of the same tube and fin surface because of earlier transition to a turbulent boundary layer and a later flow separation. Average fin efficiency for the elliptical tube, that varies with tube curvature and the height of the rectangular fin, is also higher than the uni-form fin efficiency of the round tube with a uniform height fin.

Considerations of turbulent flow and wake generation explain the higher heat transfer coefficient with smaller tubes (because of much reduced wake), and also when two or three banks of plain tubes are used as compared to one bank (the banks that follow benefit from the turbulence generated by the first bank).

With the heat transfer coefficient being a function of air velocity, heat transfer and air side pressure drop cannot

Heat Transfer and Pressure Drop With Extended Surfaces

of course be considered separately in air-cooled heat exchanger designs.

The heat transfer coefficient for a finned tube should be
expected to be of similar form to that for airflow parallel
to a uniform thickness, plate

$$\frac{hDo}{k_f} = .33 \left(\frac{C_p\mu}{k}\right)_f^{1/3} \left(\frac{DoGmax}{\mu_f}\right)^{.6} \quad (5.4)^*$$

and also to the heat transfer coefficient for airflow normal
to banks of staggered tubes.

$$\frac{hL}{k} = .0356 \left(\frac{VG}{\mu}\right)^{.8} \left(\frac{C_p\mu}{k}\right) \quad (5.5)^*$$

Where

h	the average for the surface coefficient of heat transfer
L	Length in ft in the direction of the flow of a unit width plate
k	Air thermal conductivity
V	Air velocity, in ft/hr
Do	Outside tube diameter
v	Air kinematic viscosity
x	Air thermal diffusivity
Cp	Air specific heat at constant pressure
H	Air absolute viscosity
Gmax	Air mass velocity at minimum cross-section

For rows of finned tubes of various configurations, heat
transfer and pressure drop measurements, often proprietary,
are sometimes plotted vs. Reynolds number in the form of

*Equation (5.4) in McAdams' "Heat Transmission," Third Edit-
ion, McGraw-Hill 1954, p. 272; Equation (5.5) in Jakobs'
"Heat Transfer", Vol. I, John Wiley and Sons. p.482.
Units are not consistent between the two equations.

Heat Transfer and Pressure Drop With Extended Surfaces

the Colburn heat transfer factor

$$j = \left(\frac{h}{C_p G}\right)\left(\frac{C_p \mu}{k}\right)^{2/3}$$

and a friction factor "f" or "i".

The general correlations for the heat transfer coefficient, based on the Jameson data that show "h" plotted vs. $V^{0.8}$ as nearly straight lines and for airside pressure drop as developed by Gunter-Shaw, are recommended by Kern* and are shown plotted in Figure 5.2. As considerable practice is required in their application, the worked out examples in the above reference are useful.

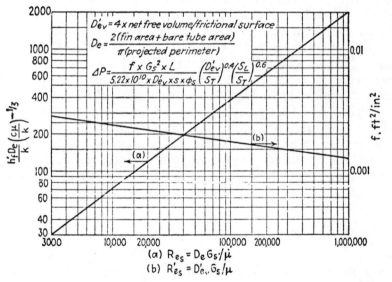

$$D'_{ev} = 4 \times \text{net free volume/frictional surface}$$

$$D_e = \frac{2(\text{fin area} + \text{bare tube area})}{\pi(\text{projected perimeter})}$$

$$\Delta P = \frac{f \times G_s^2 \times L}{5.22 \times 10^{10} \times D'_{ev} \times s \times \phi_s}\left(\frac{D'_{ev}}{S_T}\right)^{0.4}\left(\frac{S_L}{S_T}\right)^{0.6}$$

(a) $R_{es} = D_e G_s/\mu$

(b) $R'_{es} = D'_{ev} G_s/\mu$

FIGURE 5.2
Transverse-Fin Heat Transfer and Pressure Drop

* Kern, "Process Heat Transfer, McGraw-Hill, N.Y. 1950, pp. 553-559. Specific heat transfer and pressure drop data for several common fintube designs are given in Figures 11.10 and 11.11, p. 564, of "Extended Surface Heat Transfer" by Kern and Krans, McGraw-Hill, N.Y. 1972.

Heat Transfer and Pressure Drop With Extended Surfaces

As it can be shown that both the heat transfer coefficient
and the friction factor are unique functions of the air side
mass flow rate, one can be plotted directly against the
other, usually in the form of heat transfer coefficient "h"
vs. $\frac{P}{A}$ (horsepower per square foot) or vs. heat transfer surface, or sometimes vs. frontal area of the heat exchanger.
    Another form of rating heat exchanger performance and
presenting data is the so-called NTU method, based on
non-dimensional rating factors; Smith has proposed some
very general correlations for heat transfer and pressure
drop in   NTU and other non-dimensionless passage
variables.
    Equations (5.4) and (5.5) show a proportional depend-
ence of the airside film heat transfer coefficient on air
velocity raised to a fractional exponent. Correlation of
data from finned tube bundle measurements has been reported in the form of Equation (5.5), with the proportion-
ality constant varying between .2 and .4 and the velocity
(or mass flow) exponent between .5 and .8.
    Turbulent flow is a requirement in achieving reasonable
values for the airside heat transfer coefficient, as otherwise a thick boundary layer will act to insulate the fin surface from the airflow. The use of turbulence promoters or
turbulators such as ribs, slots, etc., on the fins serves to
break up the boundary layer at some sacrifice in pressure
drop and turbulent flow can be realized at lower velocities,
4-6 fps in one design.

5.3 Practical Considerations

For heat exchangers in air-cooled condensing system applications that, as discussed in the last chapter, require large
air flows at minimal pressure drops, turbulators are claimed
effective in increasing heat transfer rates at low air velocities and with low pressure drop. The effect of slots or per-
forations on the fins in increasing heat transfer rates for
the Forgo-Heller design heat exchanger is shown in
Figure 5.3; Figures 2.2a, 2.2b and 2.2c show similar
data from some early tests of an oval tube, direct
condensation air-cooled heat exchanger design.
    The effect of fouling on the inside tube surface and the
airside finned surface affect air-cooled heat exchanger performance to different degrees. Airside fouling reduces fin
airside heat transfer coefficient "h", but as seen from

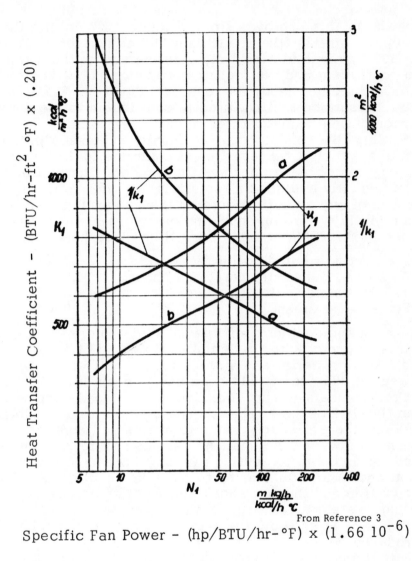

Specific Fan Power - (hp/BTU/hr-°F) x (1.66 10^{-6})

FIGURE 5.3
Performance of Heat Exchangers with Slotted-Rib Fins
(a) With Slotted-Rib Fins, (b) Fins without Slotted Ribs

Heat Transfer and Pressure Drop With Extended Surfaces

Figure 5.1, this improves fin efficiency " $\emptyset$ " so that the equivalent airside film coefficient "h" as derived in Equation 5.3 by recalculating "h'_o", as the sum of the airside film coefficient and the fouling resistance, is little affected. On the other hand, as seen from Equation (5.1), inside tube surface fouling directly contributes to the deterioration of heat transfer rates.

Fouling film coefficients are usually expressed in the form of the reciprocal resistances. Airside fouling (but not plugging of the fin surface that renders fins ineffective) is usually assigned a resistivity of .002, while in the absence of any specific figure on the inside tube fouling film resistivity for aluminum and carbon steel tubes, the values in the TEMA standards* will be given here; they are:

For heating treated boiler feedwater	.001
As above, but for temperatures under	
125°F at water velocities over 3 fps	.0005
For exhaust steam	.0005

Values for the contact coefficient, also expressed as a resistance, vary with fin design and method of attachment, tube and fin material and surface roughness, etc. Values between 1,000 and 10,000 have been measured. The contact resistance may show considerable increase during the operating life of the heat exchanger. The effect of vibrations and thermal shock on the tightness and contact resistance between tube and fins, at plant start-up, shutdown and in operation during rain or snow, have received attention. A foreign manufacturer claims negligible effect of vibration and thermal shock substantiated by extensive testing of steel tubes with aluminum fins.

Figure 5.4, showing the manufacturer's test results, is reproduced here.

It is the practice in process industry applications to allow for extra surface initially, to account for deterioration of the contact bond during service, on the basis of derating curves that take into account the average service wall temperature. Extra surface is also often allowed for to account for inside tube surface fouling.

*Standards of Tubular Exchanger Manufacturer's Association, Fifth Edition, 1968, pages 124 and 125.

Heat Transfer and Pressure Drop With Extended Surfaces

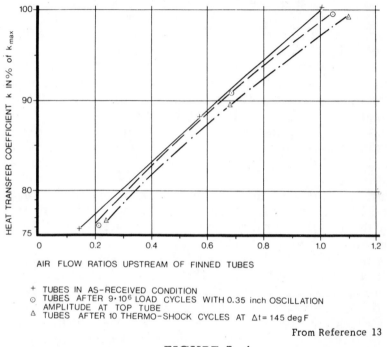

AIR FLOW RATIOS UPSTREAM OF FINNED TUBES

+ TUBES IN AS−RECEIVED CONDITION
○ TUBES AFTER 9·10⁶ LOAD CYCLES WITH 0.35 inch OSCILLATION
  AMPLITUDE AT TOP TUBE
△ TUBES AFTER 10 THERMO−SHOCK CYCLES AT Δt= 145 degF

From Reference 13

FIGURE 5.4
Effect of Vibrations and Thermal Shock
on Fin Contact Resistance

5.4 Novel Approaches to Increased Airside Heat Transfer

The importance of improving the airside heat transfer coefficient as a means for reducing air-cooled heat exchanger
size and cost and also for improving performance of power
plants with more effective air-cooled condensing systems
is obvious.

Several methods of improved cooling by increasing heat
transfer rates to the air, that have been developed for other
applications, can be of potential benefit to air-cooled heat
exchangers, although their physical applicability is some-

Heat Transfer and Pressure Drop With Extended Surfaces

times difficult to visualize and early claims are often rather exaggerated.

Artificially roughened outside tube surfaces that are being developed overseas for gas-cooled reactor heat exchangers show increased heat transfer rates, but at higher gas velocities and pressure drops than encountered with air-cooled heat exchangers.

Electrostatic cooling, often reported effective in cooling electronic equipment or pieces being welded or machined, by disturbing the boundary layer with electrical discharge between the surface and a pointed electrode, has also been discussed as a possibility. Power requirements and manufacturing feasibility, in comparison to conventional air-cooled exchangers, have been rather cheerfully understated in this case.

Various forms of heat exchanger equipment, such as fluidized bed heat exchangers that achieve high heat transfer rates in process applications, regenerative type (Ljungstrom) heat exchangers, where the two fluids come successively in contact with the same surface and are used in power plant heat recovery from the combustion gases, etc., are being studied for possible applicability to power plant waste heat rejection to ambient air. Often the physical parameters involved in some cases make such schemes appear impractical, but new suggestions of course should not be rejected a priori, and the possibility for a new method, or the applicability of a technique that appears initially farfetched, should not be excluded.

Heat Transfer and Pressure Drop With Extended Surfaces

REFERENCES

1. Gardner, K. A., Efficiency of Extended Surface; Transactions of the ASME Volume 67, November 1945.

2. Schulenberg F., Air Instead of Water for Cooling and Condensation; Chemie-Ingienieur-Technik, 25 Jahrg. 1953/Nr.10.

3. Heller, L. and Forgo, L., Operating Experience of a Power
   Plant Condenser Installation with Air Cooled Circulating
   Water and the Possibilities for Further Development; Paper 113
   G3/2, Fifth World Power Conference, Vienna 1956. (in German)

4. Schulenberg F. J. Finned Elliptical Tubes and Their Application in Air-Cooled Heat Exchangers; Transactions of the
   ASME, Journal of Engineering for Industry, May 1966.

5. Barrow, H. and Roberts, A., Flow and Heat Transfer in
   Elliptic Ducts; Heat Transfer 1970, Papers presented at
   the Fourth International Heat Transfer Conference, Paris,
   1970, Volume II, Paper FC 4.1, Elsevier Publishing Co.

6. Katz, D. L., Beatty K.O. Jr. and Foust, A. S., Heat Transfer Through Tubes with Integral Spiral Fins; Transactions of
   the ASME, November 1945.

7. Jameson, S.L., Tube Spacing in Finned-Tube Banks; Transactions of the ASME, November 1945.

8. Gunter, A.Y. and Shaw, W. A., A General Correlation of
   Friction Factors for Various Types of Surfaces in Crossflow;
   Transactions of the ASME, November 1945.

9. Smith, J.L. Jr., The Presentation of Heat-Transfer and
   Friction Factor Data for Heat Exchanger Design; ASME Paper
   No. 66-WA/HT-59.

0. Schmiechen, U., The Economic Merits of Different Types of
   Finned Tube; The Brown Boveir Review, Volume 54, No.
   10/11, October/November 1967.

1. Smith, E.C., Gunter, A.Y. and Victory, S. P. Jr., Fin Tube
   Performance; Chemical Engineering Progress, Volume 62,
   No. 7, July 1966.

2. Nevins, J. W., Finned Tube Heat Exchangers, and How to
   Calculate Surface Area Needed; Power, August 1965.

Heat Transfer and Pressure Drop With Extended Surfaces

13. Heeren H. and Holy, L., Air Cooling for Condensation and Exhaust Rejection in Large Generating Stations; Paper presented at the American Power Conference, Chicago, Illinois 1970. Reprinted in part as "Dry Cooling Eliminates Thermal Pollution   "; Combustion, October & November 1972.

14. Gardner, K.A. and Carnavos, T.C., Thermal - Contact Resistance in Finned Tubing; Transactions of the ASME Journal of Heat Transfer, November 1960.

15. Williams F. and Watts, J., The Development of Rough Surfaces with Improved Heat Transfer Performance and a Study of the Mechanisms Involved; Heat Transfer 1970, Papers presented at the Fourth International Heat Transfer Conference, Paris 1970, Volume II, Paper FC 5.5., Elsevier Publishing Co.

16. Kibler, K. G. and Wiley, R., Electrostatic Cooling; Industrial Research, April 1972.

17. Holmes, R. E. and Basham, S. J., Jr., A Dry Cooling System for Steam Power Plants; 1971 Intersociety Energy Conversion Engineering Conference Papers.

18. Andeen,   B. R. and Glicksman, L. R., Dry Cooling Towers for Cooling Plants, Report No. DSR 73047-1, Engineering Projects Laboratory, Dept. of Mechanical Engineering, M.I.T., February, 1972.

Chapter Six

MECHANICAL DRAFT AND NATURAL DRAFT COOLING
TOWERS

6.1 Draft Generated by Fans

Mechanical draft cooling towers are assemblies of heat
exchanger panels arranged so as to be evenly exposed to
the airflow created by fans, each fan usually serving two
or more tube bundle panels.

The fans can be forced draft, where the fan is forcing
the cool air through the finned tube panels, creating a
plenum of cool air between the fan and the tube panels;
or induced draft, with the fan pulling the air through the
finned tube panels, creating a slight vacuum of heated
air between the fan and the finned tubes. As in the latter
case, the air handled by the fans has a larger volume for
the same mass flow because of the heat already absorbed
and the slight vacuum in the fan's suction; the fan being
a volumetric machine, a slight penalty in fan horsepower
will be incurred. If this disadvantage is small (usually
when the temperature rise of the air through the heat exchanger panels does not exceed 30°-40°F, as is usually
the case in air-cooled condensing system application),
placing the fan after the heat exchangers in the air flow
path for induced draft is preferred as it results in better
control, less recirculation and better air flow distribution, out-balancing the theoretical penalty of the induced
fan handling lower density air.

In arranging the finned tube panels, the Vee and in-
verted Vee configurations - also referred to as A-frames -
are usually preferred, with the fan plane providing the
third side for a triangular cross section module. The
enclosed angle formed by panels is usually obtuse, the
panels being inclined to the horizontal only enough for
adequate drainage of the tubes; but designs with an acute
angle, forming the narrow Vees familiar from refining and
process industry practice, have also been proposed.

The fan and heat exchanger modules are placed atop a
structural steel or reinforced concrete platform and piped
to steam or water distribution feeders, return headers,

Mechanical Draft and Natural Draft Cooling Towers

valves, etc., with power and control cables to the fans provided, and access stairways, safety railing and maintenance provisions to make a complete installation.

The elevated structure should provide adequate access space for the free flow of air to the fans. A rough rule requires that enough free flow area be provided along the periphery of the structure below fan level to equal the plot area covered by the fan-heat exchanger modules, so that if placed adjacent to a building obstructing airflow access, the elevated structure has to be further raised to make up for the reduction in free perimeter.

As explained in the previous chapter, for air-cooled power plant applications, the large volume air flows with low pressure drops - less than 1/2" W.G. - are desirable. This requirement is best met with the high flow, low static head propeller fans. The individual blades in the larger diameter sizes are of airfoil shape, tapered and twisted so as to provide a uniform air velocity from hub to tip.

Fan diameters are determined by the volume of the air to be handled and the number of blades. Multiblade designs--four to as many as eighteen blades--are preferred, as they reduce blade loading and width and improve operating smoothness. As manufacturing economies favor large modules, fan diameters have been increasing apace; 60-ft. diameter fans are currently available.

Propeller fans are often provided with shrouds, short cylindrical stacks to guide the airflow. The fan is usually located at the narrow point of the converging diverging stack. The diverging section of the stack provides for some velocity recovery and reduction in fan horsepower with the stack also preventing heated air recirculation to the fan suction.

For a given fan configuration, noise is a function of blade tip speed. For large fans, about 30 feet in diameter and operating under 100 rpm, tip speeds around 12,000 fpm are common; tip speed is reduced and rpm increased for smaller diameter fans. The air velocity leaving the fan is in the range of 1,800 fpm. The blades for the larger diameter fans are usually hollow and lightweight, manufactured by impregnating a fiberglass cloth casing enclosed in a mold with a thermosetting resin. The blades are usually set in a cast metal hub, sometimes with provisions for changing blade pitch during

Mechanical Draft and Natural Draft Cooling Towers

operation. A circular cover on the hub spokes prevents
air recirculation.

The high speed output of the motor drives is reduced
to the lower rpm requirements of the fans through gear
reducers or multiple V-belt drives. Where reducing gears
are employed, their proper selection can reduce maintenance requirements to a negligible minimum. The American
Gear Association has published service factors for estimating reducing gear suitability. Similar criteria apply
to bearing life.

Fan ratings are developed on the basis of wind tunnel
tests by the manufacturers, as specified in the "Standard
Test Code for Air Moving Services" of the Air Moving and
Conditioning Association.

Noise reduction requirements in the approaches to the
plant depend on plant location and they differ from an industrial site to a site near a residential area. In the
latter case, the requirements can be quite subjective,
while the requirements for in-plant noise levels to protect plant personnel hearing are on the basis of sound
pressure levels at different frequencies, the higher frequencies being more objectionable. Plant personnel exposure levels are specified by the Occupational Safety
and Health Act (OSHA) of 1970.

Fan manufacturers usually can supply the results of
noise measurements from testing of the complete fan-heat
exchanger module at different distances and orientations.
Often a background sound survey of the proposed site may
prove useful. Two-speed fans may be required for sites
where background noise subsides during the night; the
lower fan speed during the night, when load also dimin-
ishes, will reduce sound pressure levels.

Control of the heat rejection process in a power plant
equipped with mechanical draft air-cooled condensers is
often accomplished by varying pitch, to the extreme of
reversing pitch and airflow for freeze protection, by two-
speed motors, or by shutting off sections of fans at low
loads or periods of low ambients. The first two methods
are far more conservant of horsepower when the plant is
operating at reduced capacity, and also useful in limiting
starting torque requirements of large fan motors.

Mechanical Draft and Natural Draft Cooling Towers

Elevation and high ambients both increase fan numbers or fan sizes and horsepower to the fans. The horsepower required to deliver the same mass flow of air, and accomplish the same heat rejection, varies inversely as the square of the air density change due to the change in altitude or in ambient temperature. In the latter case, the temperature at which the heat is rejected to the air and the steam turbine backpressure will also change.

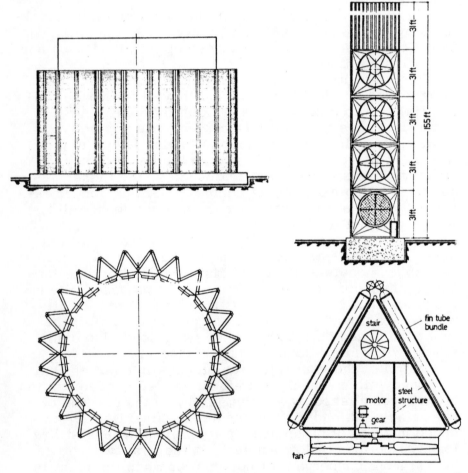

FIGURE 6.1

From Reference 9

Mechanical Draft Tower Arrangement.
Component modules are shown at right.

Mechanical Draft and Natural Draft Cooling Towers

Mechanical draft modules are marketed in various configurations and ever-increasing sizes, with induced
draft favored in the United States and forced draft in
Europe.   The standardized forced draft modules for
direct steam condensation of a European manufacturer can
be seen in the photograph inside the front cover. An in-
teresting concept proposed by the same manufacturer is
shown in Figure 6.1. In this case larger modules with
the characteristic oval plate-fin tubes, but with circulating water instead of steam for Heller system applications, are arranged with the long axes and fan planes
vertical in the periphery of a circle, the fans inducing
the airflow to the center of the fan tower. The result is
much like the vertical heat exchanger panel arrangement
along the periphery of a hyperbolic cooling tower--but
with the fans substituting for the tower's draft. Low plot
area requirements and low profile are among the advantages claimed by the manufacturer.

6.2 Hyperbolic Cooling Towers

Hyperbolic cooling towers--the familiar mammoth reinforced concrete shells used for creating natural draft when
cooling the sprayed circulating water in wet cooling tower applications--were first introduced in a Dutch colliery
in 1916; the first power plant application was in Liver-
pool in 1925. They replaced timber cooling towers of
usually rectangular cross section that were susceptible
to damage from rotting, fire and high winds.
     Their natural resistance to all three hazards accounts
for their widespread acceptance in Europe, where they
evolved in appearance through less graceful proportions
before being introduced to the United States in the 1950's.
The hyperbolic outline is due to an effort to reduce shell
surface (with resultant savings in concrete, reinforcement
and foundations) by approximating the airflow pattern in
the tower, where the incoming air turns and forms a vena
contracta prior to exiting from the slightly enlarging
stack. The shape also offers advantages against wind
forces while the thinness of the shell permitted analysis
by membrane theory, where all forces are resisted only

by compression, tension and shear in the shell. The
concrete shell, rising from a peripheral ring of X-shaped
supporting columns through which the air flows, until
recently did not exceed five inches in thickness.

The hyperbolic tower as used in air-cooled condenser
power plant applications for Rugeley and Ibbenburen has
an air flow pattern similar to the crossflow wet cooling
tower, with the finned tube heat exchanger panels arranged zig-zag fashion along the base periphery having
replaced the ring enclosing the fill. The arrangement is
shown in Figure 3.3.

As already discussed, the requirement for optimum
plant performance of large volume air flow at low pressure drop to the finned tube panels makes the hyperbolic
cooling tower, which can supply large volume flows at
low heads, ideal for air-cooled condenser power plant
applications, particularly in the larger plant ratings.

The analysis of the performance of a tower serving an
air-cooled condensing system power plant is far easier
than for the wet tower, as the only source of the draft is
from the heat rejected and the resulting chimney effect;
in the case of the wet tower, differences in air density
resulting from changes in relative humidity have also to
be taken into account.

For a given application and for similar design towers,
tower height and diameter can be traded off according to
the relation:

$$A H^{1/2} = \text{Constant} \qquad (6.1)$$

presented by Chilton in a 1952 paper on the performance
of wet hyperbolic cooling towers. In the above relation,
A is the area of a reference section parallel to the base
of the tower and H the effective tower height (above the
sill). The constant, in the wet tower case, is a function
of the water flow to the tower and the cooling range,
Merkel's cooling factor, the wet and dry bulb temperatures and the tower's fill resistance to the airflow in
velocity heads.

A similar analysis for the simplified case of a tower
serving an air-cooled power plant yields the following
relationship:

Mechanical Draft and Natural Draft Cooling Towers

$$AH^{1/2} = \frac{N^{1/2} \cdot Gt^{3/2} \cdot P^{1/2} \cdot Cp^{1/2}}{\sqrt{2} \cdot \rho^{3/2} \cdot g^{1/2} \cdot R^{1/2} \cdot Q^{1/2}} \qquad (6.1a)$$

Constant for a given application

Where:

N = Resistance to the airflow in velocity heads of the tower's heat exchanger.

Gt = Total airflow through the tower, in lbs/hr.

P = Atmospheric pressure, in lbs/ft^2.

Cp = Specific heat at constant pressure of the air .24 BTU/lb – °F for dry air.

ρ = Air density, lbs/ft^3.

g = Acceleration of gravity, ft/hr^2

R = Gas constant for Air, 53.34 ft-lbs/lb – °F

Q = Power plant heat rejection load to the tower, BTU/hr

The simple relation is useful in analyzing the effects of the various parameters on the design and performance of the tower. For example, if the effect of altitude on tower height is desired with all other factors unchanged, substituting P=proportional to ρ (as altitude affects both barometric pressure and air density), it can be seen that for constant tower diameter, height varies inversely as the square of the air density. At an elevation of 5,000 feet, air density is reduced to 85% of its sea level value and tower height has to be increased nearly 40% as compared to a tower at sea level with the same diameter, ambient temperature assumed the same at both locations.

The last reference presents an analysis based on perturbation theory of the relationship between a number of dry tower variables and scaling laws, with a discussion of optimal tower designs for various constraints.

Among other factors affecting tower selection are the space requirements along the periphery for the heat exchanger panels. Arranging the finned tube heat exchanger panels folded in Vees at the base of the tower lowers to nearly by half the required base perimeter.

Mechanical Draft and Natural Draft Cooling Towers

The height of the base opening is determined by construction economics, the preferred height of the panels making up the Vees, and the total panel surface required.

Early comparisons of hyperbolic cooling tower requirements for air-cooled condenser power plants to wet cooling towers gave a three-fold increase for the air-cooled condenser plant for the same plant performance, i.e., backpressure. Of course, the comparison is misleading as the two plants will not optimize to the same backpressure, but the large size towers required for air-cooled condenser plants in the current plant ratings dictate careful examination of soil conditions, seismic and wind design requirements at the proposed site.

The experience of Ferrybridge "C" Power Station in Great Britain, where several of a number of grouped towers collapsed in 1965, indicated that tower stresses set up by wind gusts were not wholly understood; the placing of clustered shells no closer than 1-1/2 base diameters was recommended as an interim measure along with increased shell thickness.

Earthquake loads on the towers are usually based on seismic load data for the site, as described in the Uniform Building Code, and the structural characteristics of the proposed structure.

Wind loading design criteria in the United States are developed on the basis of available isotach maps that give the frequency and intensity of past winds experienced at the site and the recommendations of the ASCE Task Committee Report, Paper No. 3269, "Wind Forces on Structures." To account for the variation of wind velocity with height, an exponential distribution law is utilized. Gust factors and the time interval required to engulf the structure are also taken into account. Similar criteria and codes are utilized in Great Britain and the Continent. The seismic and wind loading analysis of the tower are usually carried out by computer; while the ability of the structure to survive a specified calamity, e.g., the 750-year storm of the 500-year earthquake, is probabilistically reduced to express the risk factor involved during the contemplated life of the structure.

Hyperbolic cooling towers with finned-tube panels, with the Vees arranged vertical along the base periphery have experienced deteriorating performance, i.e., slight loss of condenser vacuum, with winds over 10 mph. Wind tunnel testing of the Rugeley Tower model had indicated a slight improvement in performance because of a small increase in air flow with rising wind, the inflow windward and at the sides making up for the loss of flow at the rear of the tower because of the lower pressure outside the shell. The discrepancy was found to be due to the fact that the panels exposed to the tangential wind component along the sides of the tower had the downwind panel of each Vee sheltered, while the leeward panels at the rear of the tower did not actually experience a negative pressure gradient and performed better than anticipated. Also with the heat exchanger panels located along the base periphery, the heated air tends to flow near the shell surface, with the formation of a cold air core at the center of the tower hindering draft and disturbing airflow inside the tower.

To minimize the effect of wind on tower performance and facilitate inception of draft at startup, towers are occasionally divided internally near the base into quadrants.

6.3 More Recent Developments in Natural Draft Towers

To minimize the unsatisfactory wind effects and eliminate cold air cone formation, the proprietary GKN Birwelco arrangement of the finned tube heat exchangers has been proposed, the Vee forming panels arranged horizontal with the center panels nearer to the base in the interior of the tower, as shown in Figure 6.2. With this arrangement the location of the heat transfer surface in the airflow path resembles that of a counterflow wet cooling tower. By arranging the center panels at a lower level, the air passage is reduced as the air flows into the tower under the panels. This feature is claimed to better proportion the airflow to the panels when windy conditions prevail, while the horizontal arrangement itself prevents the formation of a cold air core and tends to compensate for airflow maldistribution to the panels in cold weather; structural

Mechanical Draft and Natural Draft Cooling Towers

support cost is also reduced and ease of access improved.
The airflow to the tower with vertical at the periphery
and horizontal in the interior above the base panel configurations, indicating the favorable aerodynamic properties of the horizontal arrangement, is shown in the photo-
graphs, Figure 6.3, of the models tested for the manufacturer at the National Physical Laboratory in Great Britain

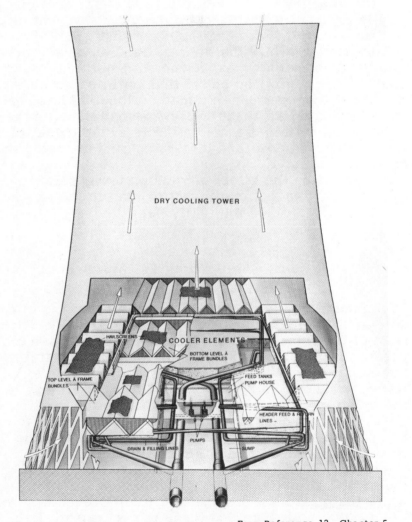

From Reference 13, Chapter 5

FIGURE 6.2
Hyperbolic Cooling Tower with Horizontal
Deployment of Heat Exchanger Panels

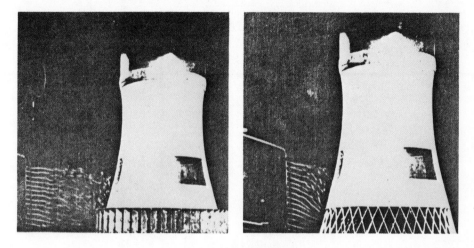

Reproduced with the Permission of GKN Birwelco Ltd.
FIGURE 6.3
Air Flow Patterns to a Tower with Vertical and
Horizontal (in the interior) Heat Exchanger Panels

The availability of more area for deploying the panels is
another argument in favor of the horizontal arrangement of
the finned tube panels inside the base of the tower. As
an example, with a 350 ft. base diameter tower, about
twice the area for arranging the panels horizontal above
the base in the interior of the tower is available, as compared to vertical arrangement in a 40 ft. opening along
the periphery of the tower. Moreover, the horizontal deployment area varies as the square of the diameter, while
the area along the perimeter is proportional only to the
first power of the diameter; the advantage in available
finned tube panel deployment area for the horizontal arrangement becomes more pronounced with increasing size
tower designs.

Another recent development is the Hoterv, Hungarian
design of an aluminum clad, structural steel frame natural
draft tower, Figure 6.4. The total weight of the tower is
reported to amount to that of the steel reinforcing alone
in a concrete hyperbolic tower. Low cost and speed of
erection should be major advantages of the design, particularly in areas where poor soil conditions will require

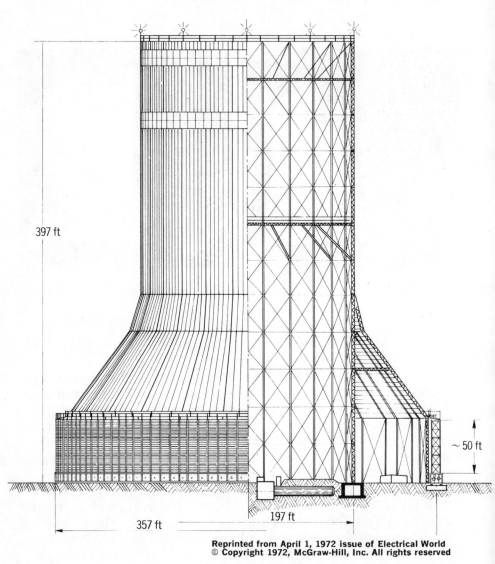

397 ft

~ 50 ft

357 ft

197 ft

FIGURE 6.4
The Hoterv Steel-Frame, Aluminum-Clad
Natural Draft Cooling Tower

major foundation expense for a hyperbolic concrete tower.
Steel framed and metal clad towers have been tried out in
the past for wet tower applications in Europe, but with
disappointing results, because of the humid and corrosive
exposure coincident with such service. Hoterv design

Mechanical Draft and Natural Draft Cooling Towers

towers have been erected for the three 220 MW units of the
Razdan, U.S.S.R. plant.

Some developments that have been discussed for wet
cooling tower applications but which, if introduced and
found successful, will also find applicability with air-
cooled condenser power plants, are attributable to the
Central Electricity Generating Board of Great Britain.
They are: the Fan Assisted Tower, where air flow to a
hyperbolic tower is enhanced by forced draft fans placed
along the base periphery; and the Ellipsoidal Tower, Figure 6.5, a large, low, semi-ellipsoidal shell that

FIGURE 6.5
Experimental Model of an Ellipsoidal
Natural Draft Cooling Tower

separates the airstream approaching the shell, creating a negative pressure at the center that induces draft.

The above developments in heat exchanger arrangements and tower configurations are characteristic of the inevitable improvements and refinements that follow the introduction of a new technology, past its early stages and coming into its own; where preconceptions and established designs in related fields have to be re-examined and adjusted to accommodate and benefit the new technology.

6.4 Relative Advantages of Natural Draft Towers and Mechanical Draft

As plant sizes increase the natural draft tower has the advantage over mechanical draft for moving the large volumes of air at the low-pressure drops required. The economics of increasing tower size and distributing the decreasing incremental tower cost over the greater generating capability benefit the natural draft tower as compared to proportionally increasing the number of fans and associated power requirements.

Mechanical draft is favored where fuel costs are low and the auxiliary power usage for operating the tower's fans--often reaching 3 to 4 % of the plant's output--can be had at low cost.

A low plant capacity factor, i.e., operation for a limited number of hours each day or extended part load operation, also favors mechanical draft selection for an air-cooled condenser power plant.

For the small plant sizes, perhaps below 300 MW, depending, of course, on design back pressure and ambient temperature, the mechanical draft tower can sometimes successfully compete with the natural draft tower even in high fuel cost areas, as the first cost of a natural draft tower cannot be easily defrayed with a smaller plant.

Other factors, such as siting considerations, may also favor the mechanical draft tower, i.e., where low load-bearing capability of the soil or prevailing high winds require extensive foundation work or additional reinforcing of the tower, or when a requirement for a low profile for a plant near an urban center makes the natural tower's bulk undesirable.

Mechanical Draft and Natural Draft Cooling Towers

REFERENCES

1. McKelvey K.K. and Brooke, M., The Industrial Cooling Tower; Elsevier Publishing Co., 1959.

2. Cooling Tower Fundamentals and Application Principles; The Marley Company, Kansas City, Missouri.

3. Rish, R.F. and Steel, T.F., Design and Selection of Hyperbolic Cooling Towers; Journal of the Power Division, Proceedings of the American Society of Civil Engineers, October 1959.

4. Harris, P.J., Billington, M.J. and Wingrove, R.J.,; U.S. Patent 3, 519,068 assigned to GKN Birwelco Ltd., England.

5. Forgo, L., Air Cooled Condensing Equipment for Thermal and Nuclear Power Stations as a Protection of the Environment; Paper 2.3-167, The Eight World Energy Conference, Bucharest 1971.

6. Ferrybridge Cooling Towers; The Engineer (G.B.), December 24, 1965 and August 26, 1966.

7. Davidson, W.C., Tower's Cooling Doubled by Fan-Assisted Draft; Electrical World, March 25, 1968.

8. Christopher, P.J. and Forster, V.T., Rugeley Dry Cooling Tower System; Proceedings of the Institution of Mechanical Engineers, 1969-1970, Vol. 184, Pt.1, No. 11.

9. v. Cleve, H.H., Westre, W.J. and Parce, J.Y., Economics and Operating Experience with Air Cooled Condensers; Paper presented at the American Power Conference, 1971.

10. Beranek. L.L., Walsh-Healy Act Changes Spur Design-out of Noise to Protect Worker's Hearing; Power May 1970.

11. Seelbach, H. and Oran F.M., What To Do About Cooling Tower Noise; Heating Piping and Air Conditioning, June 1963.

12. Hopper, B.L. and Seebold, J.G., Sound Generation in Fans for Refinery Air Coolers; ASME Paper 72-WA/FE-42.

13. Chilton, H., Performance of Natural-Draught Water-Cooling Towers; The Proceedings of The Institution of Electrical Engineers, Part II, Volume 99, 1952.

REFERENCES (Cont'd.)

14. Moore, F.K., On the Minimum Size of Natural-Draft Dry Cooling Towers for Large Power Plants; ASME Paper 72-WA/HT-60.

15. Bodas, J., Dry Cooling Tower Uses Steel Structure; Electrical World, April 1, 1972.

Chapter Seven

THE SPRAY CONDENSER

7.1 Early Designs

The introduction of the Heller system resurrected interest in the contact, jet, or spray condenser, a piece of
equipment familiar to operators of early power plants but
not much in evidence after the 1920's.
    Used originally with steam engines and a few small
steam turbines, it effected condensation of the steam by
bringing it into direct contact with the circulating water,
usually sprayed into the exhausting steam; hence the
name contact condenser as compared to the shell and
tube surface condenser where condensate and circulating water do not come in contact. After mixing and condensation, the mixed condensate and circulating water
were discharged and lost but for a fraction that was
used occasionally to replace the lost condensate as
boiler feedwater. Requirements for boiler feedwater for
the early boilers were not stringent and the loss and
replenishment of boiler feedwater did not present a
problem.
    For geothermal applications in the Pacific Northwest
and overseas, where the power generating steam that condenses is supplied from ground wells and no need for replenishing feedwater arises, a small number of spray condensers have been built in the interim, but not to the requirements for condensate purity and without the efficient
internal arrangements required for larger Heller System
units.
    To maintain the desired vacuum in the spray conden-
ser, three basic arrangements were used:
        In the barometric type, the mixed condensate
        circulating water discharged from the spray
        chamber through the tailpipe, a long vertical
        tube with the exit submerged to provide a
        seal; the difference in liquid level between
        the spray chamber and the free surface of
        the tailpipe seal determining the desired
        vacuum in the condenser.

In the ejector condenser vacuum was maintained by the ejector action of a high velocity steam or air ejector removing the accumulating water.

Finally, in the low level jet condenser- the type of spray condenser that is reappearing-the vacuum is maintained by continuously withdrawing the mixed circulating water condensate with a low NPSH circulating water pump.

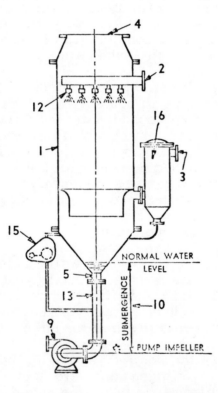

1. Condenser Shell
2. Water Inlet
3. Air-Vapor Outlet
4. Vapor Outlet
5. Water Outlet
9. Water Removal Pump

10. Water Removal Pump Submergence
12. Water Distributing Means
13. Water Pump Suction Pipe
15. Vacuum Breaker
16. Air Cooler

Reprinted from the Standards for Direct Contact Barometric and Low Level Condensers, Fourth Edition, Copyright 1957 by the Heat Exchange Institute, 122 East 42nd Street, New York, New York 10017.

FIGURE 7.1
Low Level Jet Condenser

The last two types of spray condensers provided for
more compact arrangements, as there was no space required to accommodate the thirty foot long tailpipe. A
schematic arrangement of an early low-level jet condenser is shown in Figure 7.1.

7.2 More Recent Developments

The requirements for a contact condenser for a Heller
System application in a large, modern, efficient plant
are far more stringent than those realized in the early
designs.
      The condenser must handle large volume flows of
steam and water sprays in a limited space, with good
distribution and minimum pressure drop in the steam
and sprays, achieving intimate contact and heat transfer between the steam and water while keeping irreversibilities to a minimum.
      Subcooling is to be avoided as it directly affects
cooling system capability and plant efficiency. Tightness of the condenser shell is required to avoid air in-
leakage and the condensate must be deaerated to a high
degree. Air removal, the handling of the large volume
of noncondensables, must be carried out efficiently.
Deaeration in spray condensers is discussed in Chapter 8 and subcooling is briefly discussed in Section 4.2.
      The spray condenser for the early Danube Steel
Works Plant in Hungary (16 MW) is shown in Figure 7.2.
In the period following the decision to build the first
large Heller System Plant at Rugeley, continuous efforts
to improve spray condenser performance by Heller's associates and the C.E.G.B. group suppliers, benefiting
from research into the fundamental mechanism of heat
transfer from sprays and droplets, produced considerable results and made up for forty years of inactivity
in spray condenser development.
      As reported in the Sixth World Power Conference
(1962) by Heller and Forgo, heat transfer is improved
many-fold by utilizing low pressure nozzles, the sprays
impinging on short vertical baffles to produce cascading
thin turbulent films of water with increased exposure to
the steam. The low nozzle pressure required-about 3 psi-
reduced previous atomizing power requirements nearly
threefold. Better condenser volume utilization results

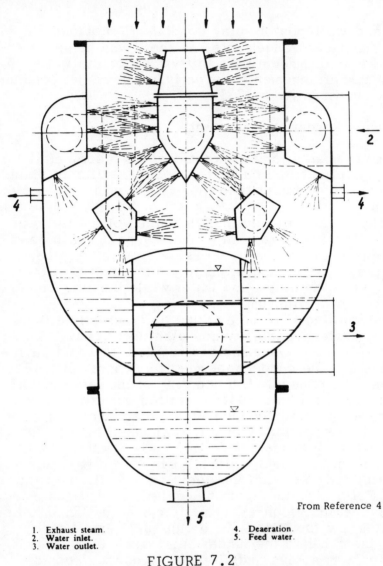

1. Exhaust steam.   4. Deaeration.
2. Water inlet.   5. Feed water.
3. Water outlet.

FIGURE 7.2
Section of the Danube Steel Works Plant
Spray Condenser

in size reduction nearly by two-thirds for the new designs, as compared to the Rugeley condenser, with
commensurate economies in fabrication, reduction of
condenser basement requirements and lower turbine
foundations. Condenser cross-section for the new
designs is reported to be determined only by the
allowable velocity of the entering steam.

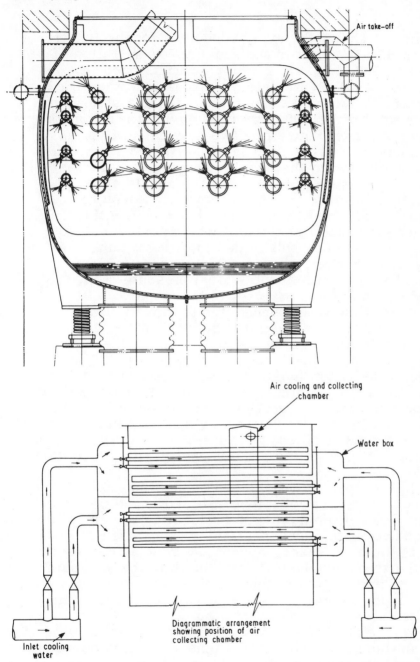

Air take-off

Air cooling and collecting
chamber

Water box

Diagrammatic arrangement
showing position of air
collecting chamber

Inlet cooling
water

From Reference 3

FIGURE 7.3
Cross Section and Sectional Elevation
of the Rugeley Spray Condenser

Figure 7.3 shows the spray condenser for the Heller
System 120 MW unit in the Rugeley Power Station of
the C.E.G.B. in Great Britain, commissioned in
December 1961. Circulating water flow to the air exchangers is about 50 times the condensing steam flow
with a total of 74,400 GPM (U.S.) withdrawn from the
spray condenser hotwell by two half-capacity pumps.
The spray condenser shell is of fabricated steel construction bolted to the exhaust housing of the three
flow low pressure turbine, with the nozzle spray
headers parallel to the turbine shaft. It occupies
approximately the same volume that a conventional
surface condenser for the unit would have occupied.
The spray nozzles, subdivided into four groups, are
large bore swirler type, producing wide angle fine
conical sprays with about 5.5 psi pressure drop. To
avoid flooding the condenser in the event of failure of
the circulating water pumps, the four hydraulically
operated butterfly valves that supply the nozzle groups
are timed to close gradually and sequentially to prevent
water hammer in the circuit. Two half-capacity head
recovery water turbines, each on a common shaft with
a circulating pump and its motor, supply about 20% of
the pump's power. One 250 lb/hr and one 150 lb/hr
steam air ejector, the latter only normally in use, remove noncondensables; the air ejectors are of the same
size as for units of the same capacity with conventional
condensers.

Initial operation disclosed subcooling up to 15°F,
inadequate removal of trapped air and impediment to
steam flow. Redesigning and relocating the air removal hoods from the condenser sides to a transverse position in the dead space between the double flow section
and the steam exhausting from the third low pressure
section, and the redistribution of a number of spray nozzle groups is reported to have eliminated subcooling and
reduced steam side pressure drop.

The condenser shown in Figure 7.4 is that of the 150
MW Ibbenbüren Power Station that went into operation in
1967 and incorporated some of the above improvements.
It also included an additional feature in its central deaerating section that exclusively serves the condensate
flow that is branched off downstream of the circulating
water pumps and returned for further deaeration.

The 2600 low-pressure spray nozzles (3.7 psi) are

The Spray Condenser

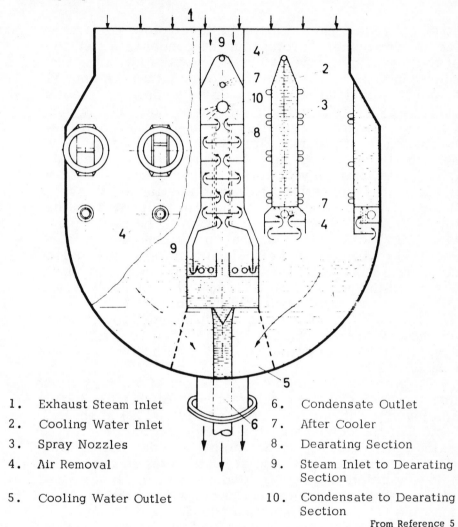

1. Exhaust Steam Inlet
2. Cooling Water Inlet
3. Spray Nozzles
4. Air Removal
5. Cooling Water Outlet

6. Condensate Outlet
7. After Cooler
8. Dearating Section
9. Steam Inlet to Dearating Section
10. Condensate to Dearating Section

From Reference 5

FIGURE 7.4
Section of the Ibbenbüren Spray Condenser

supplied from four interconnected headers. The sprays, impinging on a number of concentrically arranged baffle plates, create there continuous water films, increasing the surface area and exposure of the circulating water to the condensing steam. The noncondensables flow between the baffle plates to four extraction points, are further cooled by sprays of cool circulating water, and are discharged by water ejectors.

The deaerating achieved in the main condenser is sufficient to prevent corrosion in the circulating water lines and the cooling element tubes. The far more

stringent requirements of the boiler for low oxygen
feedwater are met by returning the condensate fraction
of the total flow, after it has been withdrawn from the
main condenser hotwell by the pumps, into the central
deaerating section of the condenser where it cascades
over a tray system and is reheated by coming into contact with steam. Originally, it was intended that the
turbine exhaust steam perform the reheating and deaerating function, but as this arrangement proved
inadequate, a supply of higher temperature steam from
the low-pressure turbine extraction performed this
function. A fifth air extraction point is located at the
top of the central deaerating section. The ratio of the
circulating water (66,000 gpm) to the condensate is
about 50 to 1. The two circulating units each consist
of a circulating water pump and motor and a recovery
water turbine mounted on the same shaft, supplying
30% of the pumping power requirements.

The spray condenser shown in Figure 7.5 is that
for the 200 MW Grootvlei, South African Power Station,
developed after an extensive test program by the German
firm M.A.N.

It provides for unobstructed entry of exhaust steam
into the center of the condenser with the circulating
spray water entering an annular header (1,2) at the
bottom of the condenser, the header feeding several
rising distributor pipes (3) supplying nozzle clusters
(4). The fine sprays reach steam saturation temperature
half way before impinging upon concentrically arranged
narrow baffles (5) releasing noncondensables as the
ensu ing thin water films cascade over baffles and
trays (6). As the steam enters the diffuser-like area
in the condenser center created by the peripherally
arranged headers of the nozzle clusters, joining the
steam descending between the nozzle sprays and
impinging baffles and reversing direction after reaching
the water surface at the bottom of the condenser, a
certain degree of recovery of the turbine exhaust steam
velocity into stagnation pressure is claimed. The
reheating of the condensate to slightly above the
temperature corresponding to the exhaust steam pressure
results in a thermodynamic advantage to the cycle.

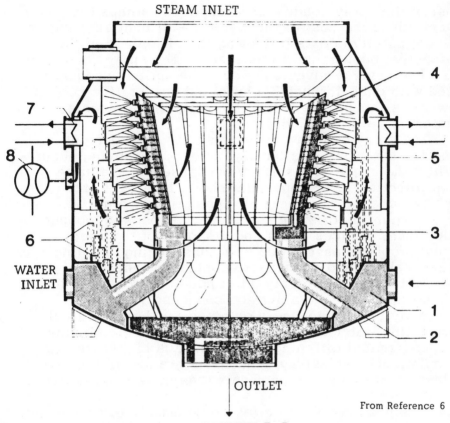

FIGURE 7.5
Section of the M.A.N. Spray Condenser

From Reference 6

Deaeration is provided by first cooling the noncondensables with cold circulating water from the upper level sprays and, after further cooling, by four generously sized vent condensers (7), to reduce air volume prior to extraction by water ejectors (8).

7.3 Contact and Surface Condensers Compared.

In the two-stage heat rejection process of the Heller system - from the working fluid (condensate) to the circulating water and from the circulating water to the air - the spray condenser is by far the least expensive part of the heat transfer equipment. The cost of the spray condenser amounts to perhaps one-twentieth of the cost of thé water-to-air heat exchangers, the comparative costs reflecting the efficiency of the respective heat transfer processes. The spray condenser also has several advantages over the surface condenser used with once-through systems and evaporative cooling towers, although the cost of the complete air-cooled system is, of course, higher.

The spray condenser, because of more efficient heat transfer, has smaller dimensions, occupying about a third the space of a surface condenser for the same heat rejection load; condenser shell, basement and turbine pedestal costs are all appropriately reduced.

Elimination of stainless steel or expensive copper alloy tubes reduces costs and disposes of the operational problems associated with tube fouling, cooling water leakage and condensate contamination.

As terminal difference is eliminated, compared to a minimum of 5°F for a surface condenser, a correspondingly better vacuum is obtained for the same final circulating water temperature. Because of the elimination of terminal difference and usually a higher rise in circulating water, spray condensers are also far more effective in series arrangements than surface condensers.

7.4 The Advantages of Series Arrangements of Condensers

In surface condensers, the benefits of series arrangements for multi-pressure operation were recognized in the 1960's with the introduction of the large multiple exhaust units.

The inidividual surface condensers or condenser sections, each usually serving two exhaust ends of a four-flow or six-flow low-pressure turbine, are operating at slightly different condensing pressures as the circulating water flowing through them in series enters each successive section at progressively higher temperatures on account of the heat picked up in the prior section.

For a two-pressure condenser the reduction in condensing temperature over a single-pressure condenser with the same surface, the same circulating water flow,

and condensing the same amount of steam is given by
the equation:

$$\Delta t_s = \frac{1}{2}\left[\left(T_d + \frac{T_r}{2}\right) - \sqrt{T_d^2 + T_r \cdot T_d}\right] \qquad (7.1)$$

Where:

Δt_s The difference between the condensing
temperature in the single-pressure
condenser and the average of the condensing temperatures in the two sections of the two-pressure condenser.

T_r The temperature rise in the single-
pressure surface condenser.

T_d The terminal temperature difference.

    The cycle improvement in performance, due to the
reduction of the steam condensing temperature in the
multi-pressure condenser, is due to the fact that heat
is transferred from the steam to the circulating water in
a more reversible fashion than in a single-pressure
condenser.

    The above equation does not include an additional
small improvement in performance realized if the condensate from the lower pressure (cold) condenser section
is reheated to the condensate tempera ure of the higher
pressure (hot) condenser section, rather than having the
two condensate streams mix. The higher condensate
temperature entering the lowest pressure feedwater
heater proportionately reduces the required steam extraction, resulting in extra power generation as additional
steam flows through the turbine section between the last
extraction point and the condenser.

    The improvement in cycle efficiency resulting from
the reduction of the steam condensing temperature in a
two-pressure condenser is proportional to the difference
in backpressure between the single-pressure condenser
and the average of the backpressures corresponding to
the two condensing temperatures in the two-pressure
condenser, as can be surmised from the manufacturer's

heat rate correction curves for small variations in back-pressure. The considerable increase in the exhaust loss of the cold turbine exhaust end is not quite compensated for by the reduction in the exhaust loss of an equal size hot exhaust end and the improvement in efficiency in practice is less than indicated by Equation 7.1.

As also can be surmised from the equation, improvements in performance are greater for higher temperature rises of the circulating water and smaller terminal differences. The latter are held to at least 5°F in surface condensers for valid reasons, as recommended by the Heat Exchange Institute, while circulating water rises are small where there is abundance of circulating water. As a result, there is little actual improvement in performance through multi-pressure surface condensers with once-through cooling. The small extra cost for segmenting a long condenser into differing pressure zones, with slight differences in hotwell elevations to cause the reheating of the condensate by gravity flow to the higher pressure section without the need of a separate condensate pump, has led to the adoption of the multi-pressure surface condenser for several large installations. The slight improvement in performance is usually traded off for a reduction in condenser surface.

It has also been recognized that greater improvements in performance were to be realized with multi-pressure surface condensers serving evaporative cooling towers, where the smaller circulating water flow results in a higher temperature rise, economically appropriate to power plants with cooling towers.

7.5 Cascaded Spray Condensers

With cascaded spray condensers the benefits of reducing irreversibilities in the heat transfer between steam and water and reheating of the condensate are incorporated in the simple expression:

$$\Delta t_s = T_r \cdot \frac{n-1}{2n} \qquad (7.2)$$

Where:

Δt_s As before, is the reduction in the
              average steam condensing temperature

T_r The circulating water temperature
              rise and, in this case, the cooling
              water range in the air-cooled heat
              exchangers as well.

n The number of individual spray con-
              densers in series, operating at
              different condensing steam pressures.

The terminal temperature difference T_d is absent
from this expression, as T_d should be close to zero
for an adequately designed jet condenser.

The maximum benefit that can be realized, with a
large number of jet condenser stages serving an equal
number of exhaust ends will result in halving the effective
cooling range. The effect on plant efficiency and air-
cooled system size can be seen by referring to Figure
4.7. The "effective ITD" that determines turbine backpressure and plant efficiency is reduced by the reduction
in the equivalent cooling range by connecting the spray
condensers in series. Conversely for the same average
condensing temperature as a single spray condenser
installation air-cooled system, size can be reduced.
In one actual plant design it has been found that
the series connection of jet condensers results in either
an increase of the thermal efficiency by 4-7% or in a
reduction of the dry cooling system size by 15-20% for
the same performance as with a single-pressure condenser.
A large number of stages is not required in practice
as it can be seen from the plot of the equation in
Figure 7.6 that half the maximum benefit can be realized
with two condenser stages, two-thirds with three stages,
and three-quarters with four stages; they are respec
tively one-quarter, one-third, and three-eighths of the
cooling water range. Subsequent reductions by increasing the number of condenser stages are small.

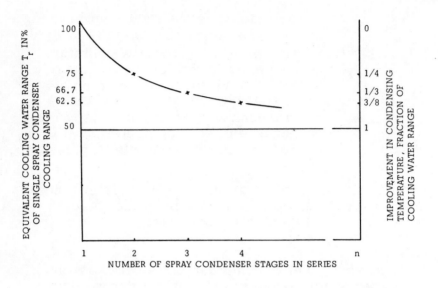

FIGURE 7.6

Reduction in Equivalent Cooling Water Range
with Cascaded Spray Condensers

Another reason that multi-pressure applications
are more effective with jet condensers than with surface condensers is that exhaust losses are smaller in
magnitude for steam turbines condensing to dry cooling
systems, and the smaller increases in the exhaust loss
of the colder sections do not distract from overall performance.

As the spray condenser is a fairly simple and inexpensive piece of equipment, there is little financial
penalty in providing two or three smaller shells rather
than a single shell of larger dimensions, but for the
cost of the additional condensate piping. A possible
disadvantage--individual condensate pumps after each
spray condenser--can be circumvented if the cascading
spray condensers are set at different elevations,
Figure 7.7, as proposed by Professor Heller. The condensate circuit between the condensers forms a water
leg, loop seal fashion, whose height accounts for the
pressure difference between the condensers and the
pressure drop in the spray nozzles.

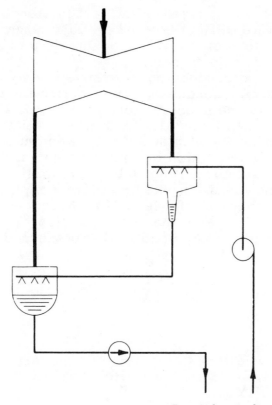

FIGURE 7.7

Jet condensers in series, the water flow
passed on by gravity from one to the other
being controlled by the water level prevailing in the lower extension of the upper
condenser.

The arrangement is made possible by the development of the low-pressure drop film spray nozzles by
Professor Heller; as with conventional high-pressure
drop spray nozzles, the required vertical distance
between condensers would have made the arrangement
impractical.

Series connection of air-cooled heat exchangers
on the air side are also possible, the heated air
successively flowing over tubes carrying warmer water,
or steam condensing at higher pressures. Such

arrangements, discussed in the last reference, present
some hardware difficulties and are also compromised
by the need to keep air side pressure drop to a minimum,
as discussed in Section 4.2.

Air-cooled condensing systems are also proposed
utilizing surface condensers and offering the obvious
advantages of the separate condensate and circulating
water circuits that simplify feedwater chemistry (particularly important with supercritical and nuclear cycles),
assure freeze protection by including an antifreeze
solution in the circulating water, and facilitate quick
and frequent starts for peaking/cycling applications.
Surface condensers offer less of an advantage in multi-
pressure arrangements than spray condensers and the
two types are compared and further discussed in Appendix D.

REFERENCES

1. Standards for Direct Contact Barometric and Low Level
    Condensers, Fourth Edition ; Heat Exchange Institute,
    New York, N.Y.

2. Christopher, P.S., The Dry Cooling Tower System of
    the Rugeley Power Station of the Central Electricity
    Generating Board; English Electric Journal, Vol. 20,
    No. 1, Jan/Feb. 1965.

3. Christopher, P.S. and Forster, V.T., Rugeley Dry
    Cooling Tower System; Proceedings of the Institution
    of Mechanical Engineers, 1969-1970, Vol. 184,
    Pt. 1, No. 11.

4. Heller, L. and Forgo, L., Recent Operational Ex-
    periences Concerning the "Heller System" of Air
    Condensation for Power Plants. Latest Results of
    Developments. Sixth World Power Conference 1962.
    Paper 154, III. $3_2/8$.

5. Scherf, O., The Air Cooled Condensing Plant for
    the 150 MW Unit of the Preussag Power Plant at
    Ibbenburen (in German); Energie und Technic,
    July 1969.

REFERENCES (Cont'd.)

5.   Heeren, H. and Holly, L., Air Cooling for Condensation
     and Exhaust Heat Rejection in Large Generating Stations;
     Proceedings of the American Power Conference, Vol. 32,
     1970.

7.   Weinberg, S., Heat Transfer to Low Pressure Sprays of
     Water in a Steam Atmosphere; The Institution of
     Mechanical Engineers Proceedings (B), 1952-1953.

8.   Heller, L., Efforts to Combat Irreversibilities (in German);
     Energietechnic, November 1954.

9.   Heller, L., Series Connection of Jet Condensers on
     the Cooling Water Side; VII World Power Conference,
     Section C1, Paper 149, Moscow, 1968.

0.   Berman, L.D., Some Problems of Designing Condensers
     for Large Turbines; Thermal Engineering (Teploenergetica),
     HRVA Translations, December 1965. Also in Combustion,
     January 1967.

1.   Brazier, P.H., The Economic Advantages and Limita-
     tions of Multi-Stage Condensers and Contact and
     Multi-Part Feedheaters; Turbine-Generator Engineering,
     Published by AEI Turbine-Generators Ltd., England.

2.   Palmer, W. E. and Miller, E.H., Why Multi-Pressure
     Surface Condenser Operation; Proceedings of the
     American Power Conference, Volume XXVII, 1965.

3.   Peake, C. C. and Coit, R.L., New Designs for Large
     Condensers Raise Efficiency, Reduce Corrosion and
     Facilitate Installation, Part 1; Power Engineering,
     December 1966.

4.   Budnyatskii, et al., Engineering and Economic Reasons
     for the Use of the Heller Dry Cooling Tower System in
     the USSR Power Industry; Thermal Engineering (Teplo-
     energetica), HRVA Translations, November 1969.
     Also in Combustion, April, 1971.

5.   Kelp, F., The Multistage Arrangement of the Conden-
     sation in Air-Cooled Steam Power Stations (in German);
     Brennstoff-Warme-Kraft, Volume 24, No. 9, September,
     1972.

Chapter Eight

DEAERATION AND FEEDWATER CHEMISTRY WITH AIR-COOLED CONDENSING SYSTEMS

8.1 The Need for Effective Condensate Deaeration in Power Plants

The problems of adequate deaeration and proper feedwater chemistry followed the introduction of the first power-house boilers and the subject had developed considerable literature by the turn of the century.

The condenser, whether it is a surface condenser, the direct system's steam-to-air heat exchanger or an indirect system's jet condenser with the associated water-to-air heat exchangers, is vital to deaeration and a possible source of condensate contamination.

Dissolved gases in the power plant's working fluid have a two-fold effect.

> Non-condensables in the condenser steam adversely effect cycle performance as they reduce the steam condensing coefficient and condenser effectiveness, raising turbine back pressure and heat sink temperature. It has been estimated that the presence of only one percent of air by volume may reduce the condensing coefficient of steam by 50 percent.*

> Dissolved oxygen and carbon dioxide in the condensate water increase iron solubility. Iron from the boiler and feedwater piping goes into solution; the process stops, as the reaction reaches equilibrium, when the ferrous hydroxide formed increases the alkalinity of the solution. Dissolved oxygen in the condensate, by precipitating ferric hydroxide, allows the reaction to proceed further. Dissolved carbon dioxide, by forming carbonic acid, lowers the alkalinity of the solution and again increases the amount of the dissolved iron.

*Kern: Process Heat Transfer, Chapter 12, "Condensation of Single Vapors."

Deaeration and Feedwater Chemistry With Air-Cooled
Condensing Systems

Pure water ionizes into equal number hydrogen (H^+) and hydroxyl (OH^-) ions. The product of their concentrations at
77F, in gram equivalents per liter, is 10^{-14}. Predominance
of hydrogen ions makes the solution acidic, while predominance of hydroxyl ions designates an alkaline condition.

Alkalinity of the water is measured on the logarithmic,
pH scale. A solution's pH is the logarithm of the reciprocal of the hydrogen ion concentration in grams per liter.
The extremely low hydrogen ion concentration of 10^{-14}
grams per liter has a pH of 14 and is highly alkaline. A
pH of 7, 10^{-7} grams each of hydrogen and hydroxyl ions
per liter, designates a neutral solution; a solution of
lower pH is acidic.

Removal of entrained and dissolved air from the condensate is accomplished in the condenser and in the deaerator. Both these pieces of equipment bring the condensate, broken up into small droplets or thin films to
increase its surface, into intimate contact with steam
that also heats the condensate to saturation temperature
at the respective pressures, at which temperature no
gases can remain in solution. The condenser and the
deaerator are equipped for air removal. Sodium sulfite
and hydrazine treatment have been used to remove the
remaining oxygen from the condensate while ammonia,
morpholine, and cyclohexamine control alkalinity.

The complex subject of feedwater treatment includes
not only deaeration, oxygen removal, and pH control --
touched upon above -- but also control of dissolved
solids, demineralization, and silica and copper carry-over to the turbine. These last items do not present
problems peculiar only to air-cooled condenser power
plants and are not further mentioned in this chapter.
Feedwater treatment is discussed in some detail in the
first three references.

The chief source of oxygen contamination of the condensate is air infiltration into power plant equipment
normally under vacuum, such as the low pressure end
of the steam turbine, the condenser, pump suctions, and
the first feedwater heater. The problem is accentuated
with frequent plant shutdown as oxygen dissolves readily
into subcooled condensate.

Deaeration and Feedwater Chemistry With Air-Cooled
Condensing Systems

8.2 Deaeration Experience with Large Air-Cooled
      Condensing Systems

The direct steam-to-air condenser has a large volume
under vacuum and the probability of air infiltration is
proportionally increased, but experience with existing
installations has been excellent and the long-term air-
tightness of the design has been proven. Evaluation of
the direct system's condenser, establishing vacuum for
plant start-up after air was allowed into the system following repairs or a long shutdown, has been lengthy; but
it was found that the process can be accelerated by venting the condenser and purging with steam, which improves
evacuation and air removal equipment performance.

The Heller System does not have excessive equipment
volume under vacuum as the jet condenser is of smaller
dimensions than a comparable surface condenser and
without the latter's multiplicity of tubes. Also, the air-
cooled heat exchanger tubes are under pressure to prevent
air infiltration in contrast to the direct system's steam-
to-air heat exchanger tubes that are under vacuum; but
the large volume of the mixed condensate-circulating
water, much of it subcooled, is difficult to deaerate and
maintain oxygen free. Another problem arises as a high
pH requirement for oxygen control is incompatible with
the usual aluminum tubes of the Heller System's heat
exchangers, since it increases aluminum solubility.

The Rugeley Air-Cooled Condenser Plant has experienced no corrosion from oxygen in the condensate; condenser deaeration and air removal has been adequate
after the jet condenser design was modified, as discussed
in Section 7.2. Dissolved oxygen in the condensate leaving the jet condenser is at 100 to 300 ppb and further
reduced to 20 ppb in the deaerator; remaining oxygen is
eliminated by chemical treatment before the boiler.

At Ibbenburen the mixed condensate-circulating water
is separated into two streams prior to leaving the jet
condenser. The condensate stream is further deaerated
in an inner condenser, as also discussed in Section 7.2,
and dissolved oxygen concentration is reduced to 10-20
ppb; while in the circulating water stream the dissolved
oxygen remains high at 100-300 ppb.

Deaeration and Feedwater Chemistry With Air-Cooled
Condensing Systems

Evacuation of the Heller System presents no difficulty,
with air removal equipment sized to the requirements usual
for surface condensers. At Rugeley, a vacuum is drawn
in the condenser in about 10 minutes.

Figure 8.1 shows the ability of another jet condenser
design, that of the German firm M.A.N., to deaerate a
fresh charge of oxygen-saturated water. Reduction of
dissolved oxygen in the condensate to acceptable levels
after a normal shutdown should be a matter of a few minutes. The M.A.N. jet condenser has been installed at
the Grootvlei Plant in South Africa.

8.3 Feedwater Chemistry Problems with the Indirect
and Direct Air-Cooled Condensing Systems

The complexities of feedwater treatment introduced by the
large tube surface of air-cooled condensers were recognized early.

Heller considered the problem of aluminum solubility
in the alkaline feedwater required for oxygen removal.
He reported low aluminum solubility at a pH of 8, with
total dissolved iron, aluminum, and copper less than 10
ppb.

Another problem arises with copper ions in the condensate, usually present because of copper bearing alloys in
feedwater heater tubing, pump bearings, and various
other copper-bearing equipment parts with which condensate comes into contact. In the presence of copper ions
in the condensate, an electrochemical reaction takes
place accelerating aluminum solubility. Heller also reported no difficulty with copper in spite of the fact that
the aluminum tubed air-cooled condenser plant under
study shared copper-bearing condensate with surface
condenser-cooled units.

Rugeley experienced no feedwater chemistry problems in spite of comparatively high condensate alkalinity. Circulating water piping has been internally
lined with a plastic coating. The pH at 8.8 to 9 has
been controlled by slug-dosing the boiler feedwater
with morpholine, keeping the boiler water pH at 10.
Rugeley also occasionally shared condensate with copper alloy tubed surface condensate units. At Rugeley,

Deaeration and Feedwater Chemistry With Air-Cooled
Condensing Systems

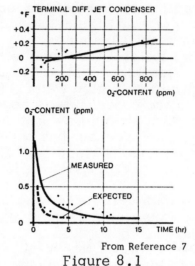

From Reference 7

Figure 8.1
Oxygen Content and Terminal Difference
During Start-up for the M.A.N. Spray Condenser

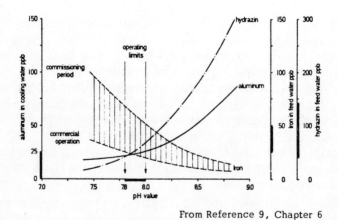

From Reference 9, Chapter 6

Figure 8.2
Effect of the pH Value (Hydrazine Content)
on Aluminum and Iron in the
Condensate and the Circulating Water
at the Ibbenbüren Spray Condenser

aluminum in the condensate has varied between 10 to 20
ppb while soluble iron remained at 20 ppb. No deposits
in the boiler or the turbine have been experienced although the plant has operated in cycling service, which
is more severe for corrosion.

The Ibbenbüren Plant, on the contrary, has experienced
a considerable variety of feedwater chemistry problems.
With the pH of the condensate maintained at 8.5 and
above by hydrazine injection, aluminum content increased
sharply. More precise pH control between 7.8 and 8.0
reduced soluble aluminum to 20 ppb and, under the cir-
cumstances, a life of over 20 years for the air-cooled
exchanger tubes is predicted. Aluminum is removed from
the condensate to the boiler by a full flow filter made up
of granular, cation exchange and mixed bed sections,
after which hydrazine is injected to the condensate. The
low iron content of the condensate, 10 ppb, and the absence of corrosion have been attributed to a possible
cathodic-type protection from the aluminum air-cooled
tubes. Figure 8.2 summarizes observed concentrations
of aluminum and iron in the condensate at Ibbenburen
versus hydrazine-controlled pH levels.

The Ibbenbüren Plant has operated baseload with few
interruptions and with particular attention to the elimination of all copper from the condensate circuit. The higher
steam pressure and the condensate purity requirements
of the once-through boiler design can perhaps explain the
variety of their experience, as compared to the lack of
feedwater chemistry problems at Rugeley. Feedwater
treatment problems and experience also tend to show
pecularities from one plant to the next. The relevant
experience in the United States from a test installation
of aluminum tubes in a surface condenser at a once-
through cooled plant are given in the last reference.

The direct air-cooled steam condenser problems with
feedwater chemistry are due to the high iron content of
the condensate because of the large surface of the carbon
steel tubes favored with this system. Full flow filters
to remove iron are required and, prior to plant start-up,
several days of condensate flow through the filters is
needed to remove accumulated iron oxides.

Deaeration and Feedwater Chemistry With Air-Cooled
Condensing Systems

For the severe feedwater chemistry requirements of once-
through boilers, the problems arising from cycling service,
and those associated with nuclear plans, the air-cooled
system utilizing a surface condenser instead of a jet condenser with separate cooling water and condensate circuits, as discussed at the end of Section 7.5, may prove
attractive as it reduces manyfold the amount of condensate
to be treated; and by removing the large surface of the
air-cooled element tubes from the condensate circuit, it
simplifies condensate treatment.

Deaeration and Feedwater Chemistry With Air-Cooled
Condensing Systems

REFERENCES

1. Straub, F.G., Proven Practices in Chemical Control for
Steam Power Plants; Transactions of the ASME, Journal
of Engineering for Power, July 1967.

2. Betz Handbook of Industrial Water Conditioning, Sixth
Edition, Philadelphia, 1962.

3. Combustion Engineering, Chapter 8: Water Technology,
Revised Edition, New York, 1962.

4. Heller, L. and Forgo, W., Recent Operational Experiences Concerning the 'Heller System' of Air Condensation
for Power Plants. Latest Results of Developments;
Paper 154 III. $3_2/8$, Sixth World Power Conference,
Melbourne, 1962.

5. Christopher, P.J. and Forster, V.T., Rugeley Dry Cooling Tower System; Proceedings of the Institution of
Mechanical Engineers, Vol. 184, Pt. 1, No. 11.

6. Scherf, O., The Air-Cooled Condensing Plant for the 150
MW Unit of the Preussag Power Plant at Ibbenburen (in
German) Energie and Technic, July 1969.

7. Heeren, H. and Holly, L., Air Cooling for Condensation
and Exhaust Heat Rejection in Large Generating Stations;
Paper presented at the American Power Conference, Chicago 1970.

8. Hagewood, B.T., Klein, H.A. and Voyles, D.E., The
Control of Internal Corrosion in High-Pressure Peaking
Units; Proceedings of the American Power Conference,
Volume 30, 1968.

9. Riedel, W.L., The Control of Metal Pickup in Cycles
with Steel Tube Feedwater Heaters; Proceedings of the
American Power Conference, Volume 26, 1964.

Deaeration and Feedwater Chemistry With Air-Cooled
Condensing Systems

REFERENCES (Cont'd.)

10. Simon, D.E., Copper Turbine Deposits Can be Controlled; Power, February 1968.

11. Pollock, W.A., A Report on a Plant Installation of Aluminum Condenser Tubes; Proceedings of the American
Power Conference, Volume XIX, 1957.

Chapter Nine

THERMAL CYCLE ARRANGEMENTS

9.1 Considerations Affecting Cycle Selection

The design of power plants with dry cooling systems for heat rejection requires a reassessment of cycle arrangements, plant optimization procedures and equipment selection priorities.

The mode of heat rejection from the power plant, the requirement for heat transfer to air, a medium of fluctuating temperature and poor heat transfer ability that necessitates large and expensive heat transfer areas, establishes a set of guidelines useful in the consideration of appropriate thermal cycle arrangements.

A. As the receiving air temperature varies, the turbine exhaust end is required to accommodate and perform well under variable back-pressure that results in wide variations in exhaust steam volume flow.

B. As high summer temperatures and resulting high backpressures will cause reduction in plant output during periods when demand for power is usually high, means should be considered for maintaining plant output by shutting off feedwater heaters or by increasing the steam supplied by the boiler. The increased steam flow will also partially offset the decrease in the exhaust volume flow resulting from the high backpressures.

C. The high cost of the dry cooling system and the proportional cost of fan power or draft inducing structure required, makes any schemes that will reduce the amount of heat rejected worthy of detailed consideration. The obvious improvement in efficiency associated with higher steam temperatures will not be discussed here because of the well-known material problems associated with the higher steam temperatures. Combinations of the steam cycle with other methods of energy conversion, such as thermionic or MHD topping cycles or combined gas turbine-steam turbine cycles, for improved overall efficiency and reduced heat rejection will also not be considered, as any schemes for reduced heat rejection in the steam cycle will also apply to such arrangements.

Other obvious means for improving steam cycle efficiency are the increase in the number of feedwater heaters by one or two over the number of five or six usually found in high efficiency base load steam plants, and the selection of a second reheat.

D. Heat transferred across the dry cooling system heat exchangers (the quantity of heat rejected) other factors being equal, is roughly proportional to the temperature difference of the fluids involved, steam or water and air. Ambient air temperature cannot be controlled, although spraying water into the air stream in a fashion similar to cooling the intake air of gas turbines has been suggested as a means for cooling the air and also improving draft during adverse ambients. Thus any schemes for increasing the effective temperature of the other medium without affecting the performance of the cycle, will result in reduction of dry cooling equipment size and proportional savings or, for the same dry cooling equipment size, in improved overall performance.

The last statement appears to be a contradiction in the light of the second law of thermodynamics--improving performance while increasing heat rejection temperature-- but that this is not the case is made clear in the discussion of staged (or multipressure) spray condensers.

E. As expansion line end point enthalpy, exhaust loss and temperature of the exhaust steam, and through them dry cooling system selection, depends on initial steam conditions, number of steam reheat stages and the associated pressure drops, the selection of initial steam pressure and reheat conditions will affect dry cooling system size and cost.

F. For plants with low load factors, possible sacrifice of plant efficiency should be considered for lower overall plant cost through the selection of a smaller dry cooling system for higher backpressures that will result in higher temperature differences between steam or water and air and higher heat transfer rates across the dry cooling system heat exchangers.

Of course, the same capital cost savings approach should apply to the selection of the rest of the plant equipment, as discussed in detail in the chapter, "Power Plants for Peaking/Cycling."

G. Aside from increased extraction to a larger number of feedwater heaters for improved efficiency discussed in Item C, any other means for reducing steam exhaust flow should be considered.

Extraction for steam driven auxiliaries, i.e., for boiler feed pumps and the large fans that reduce exhaust end loading and eliminate driver motor electrical losses, will only have a secondary influence in reducing heat rejected, as the steam used by the auxiliaries' drivers also has to be condensed by air cooling.

Extracting and condensing steam for combustion air heating in lieu of a regenerative air heater, as distinguished from limited preheating for raising average cold end temperature to avoid corrosion in the air heater, results in considerable reduction of heat rejected by the dry cooling system. Although in this case combustion gases leave the boiler at considerably higher temperatures, as the regenerative air heater will have been eliminated while the boiler economizer surface will have been increased somewhat, overall cycle thermal performance shows little change and considerable savings in dry cooling system and other equipment costs result.

H. The perennial question of fluids other than steam for the power cycle should again be considered. Little improvement in heat exchangers can result because of fluid properties alone, as the overall coefficient of heat transfer is controlled by the air side film coefficient, but other fluids offer the prospect of eliminating the large variations in exhaust volume flow accompanying backpressure changes that are a problem with steam.

Also, the prospects offered by the freons and ammonia for eliminating the freezing problems associated with steam at low ambients, guarantees that refrigerants be considered as working fluids for the power cycle. This topic will be discussed in a subsequent chapter.

How these guidelines affect thermal cycle arrangements will be considered in some detail in this chapter, while questions of priorities in large equipment selection and overall plant optimization will be discussed in the next chapter,

9.2 Effect of Initial Pressure and Double Reheat.

The effect of initial pressure and double reheat on heat
rate, together with that of recently proposed higher efficiency cycle, is shown in Figure 9.1. Raising initial
pressure at constant steam temperature decreases the
enthalpy of the main steam to the turbine but nonetheless improves efficiency, as saturation temperature is
raised and with it the amount of heat absorbed at a
higher temperature. Also the advantage derived from
feedwater heating is largely dependent on initial pressure, a higher pressure corresponding to a higher optimum temperature of regenerative feedwater heating,
which improves efficiency, as the heat added from an
external source is at a higher average temperature.
 Raising steam pressure increases the stresses in
the boiler and in the main and reheat steam piping but
also reduces the specific volume of steam and the required boiler tube and steam pipe sizes. The higher
equipment costs of a 3500 psig (supercritical) plant
relative to a 2400 psig once through and 2400 psig,
one and two drum boiler power plants, are shown in
Table 9.1, taken from the first reference. The average improvement of net plant heat rate at the higher
pressure is reported as 1.6%, (9000 vs. 8860 Btu/kWhr.)
 The effect of reheat parameters and double reheat on
plant costs and efficiency is rather complex. Reheat
was introduced in the mid-1920's primarily to reduce
exhaust steam moisture and the erosion of the last
turbine stages at the steam conditions common to that
period; improving initial steam conditions thereafter
limited reheat use until the 1940's. Reheat improves
efficiency by raising the average temperature of the
heat input to the cycle. The same applies to double
reheat, with additional benefits derived from increased initial steam pressure, higher first reheat temperature and higher feedwater temperature, with a greater
number of feedwater heaters and increased extraction,
all usually accompanying the adoption of double reheat
for a power plant, over a single reheat design.

TABLE 9-I

COMPONENT	FUEL			
	Coal		Oil or Gas	
	1-3500 psi vs 1-2400 psi O-T	1-3500 psi vs 2-2400 psi Dr	1-3500 psi vs 1-2400 psi O-T	1-3500 psi vs 1-2400 psi Dr
Boiler	+1.00	+1.00	+0.50	+2.00
Turbine	-0.54	-0.54	-0.54	-0.54
Condenser	-0.04	-0.04	-0.04	-0.04
Feedwater heater and piping	+0.28	+0.28	+0.28	+0.28
Feedpump, drive	+0.25	+0.25	+0.25	+0.25
Steam piping	0	-1.00	0	0
Total:	+0.95	-0.05	+0.45	+1.95

Differential Costs in $/kW between 3500 psig and 2400-psig once-through (O-T) and drum-type (Dr) boilers, 800-MW capacity.
Reprinted with permission from Power, May 1970.
Copyright McGraw-Hill, Inc., 1970.

From Reference 1

For conventionally cooled plants, single reheat pressure is usually 18-23% of initial steam pressure, for double reheat the first reheat is in the range of 27-37% and the second reheat is 7-10% of the initial steam pressure. Figure 9.2 indicates the variation in efficiency with reheat parameters at the low backpressure usual with once through cooling. The analytical treatment of the choice of reheat parameters and the effect of the higher backpressures with air cooled condenser plants is rather complex; increased backpressure reduces the efficiency of reheat as a larger proportion of the heat added by reheat is rejected and the optimum pressure for the first and second reheat will be shifted to the higher values of the ranges given above.*

*The argument for higher reheat pressures at high turbine backpressures is illustrated in a simple manner in p.339 of W.J. Kearton's "Steam Theory and Practice," Seventh Edition. Pitman, London 1958.

The effect of double reheat on plant cost can best be determined by evaluating optimized complete designs in each case. A foreign manufacturer estimates a 10% increase of the turbine cost and a 5-8% increase in overall plant costs with double reheat for large conventionally cooled plants. Figure 9.3 summarizes the results of a domestic study by a consulting engineering firm, based on detailed estimating of complete plant designs.

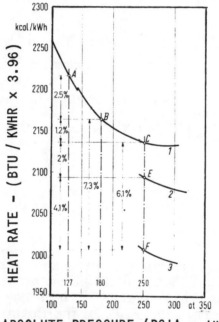

ABSOLUTE PRESSURE (PSIA x 14.7)

From Reference 4

HEAT RATE OF STEAM POWER PLANTS VS. THROTTLE PRESSURE
WITH 540° C. (1 004° F.) STEAM TEMP.

1. SINGLE REHEAT TO 540° C.

2. DOUBLE REHEAT TO 540° C.

3. DOUBLE REHEAT TO 540° C. AND USE OF THE
   SONNEFELD - ESCHER - WYSS PROCESS FOR
   FURTHER REHEATING STEAM TO 520° C.
   (968° F.) IN A STEAM TO STEAM REHEATER

FIGURE 9.1
Effect of Initial Steam Pressure
and Double Reheat on Heat Rate

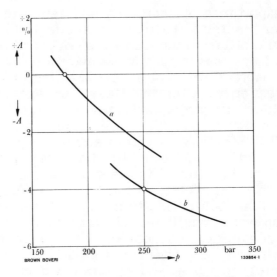

Change in specific heat consumption with double reheat

A = Change in specific heat consumption
p = Live steam pressure
a = Single reheat 540/540 °C
b = Double reheat 540/540/540 °C
Correction $A' = \pm\, 0\cdot25\%$ if steam temperature varies
by $\pm\, 10$ °C

From Reference 3

FIGURE 9.2

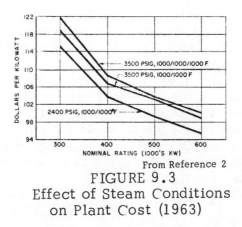

From Reference 2

FIGURE 9.3
Effect of Steam Conditions
on Plant Cost (1963)

The importance of improving efficiency in air-cooled condenser plants is greater than suggested by fuel savings alone - the criterion used in trade-offs of efficiency vs. additional equipment cost for conventionally cooled power plants. Improvements in efficiency also result in a more drastic reduction of turbine steam rate and heat rejected by the plant and corresponding reduction of air-cooler condenser size and cost. For example, a 5% improvement in a 9000 Btu/kWhr heat rate results in over 8% reduction in heat rejected.* As a result, improvements in efficiency by the adoption of double reheat or advanced steam cycles such as The Field Cycle or the Sonnefeld-Escher-Wyss Process, that require additional equipment expenditures and which were considered of marginal value when evaluated for conventionally cooled plants, may be proved worthwhile when the additional savings resulting from reducing rejected heat and air-cooled condenser costs are taken into account. The performance of the partial steam recompression, triple reheat Sonnefeld Process is shown in Figure 9.1.

9.3   Combustion Air Preheating by Extraction Steam

Preheating of combustion air by extraction steam, above that required for raising the air heater cold end temperature for corrosion prevention, has received considerable attention abroad. In the original CESAS cycle (the title is an acronym from the French equivalent of "Preheating Air by Extraction Steam and Preheating Water by the Gases", developed by Societe' Francais de Constructions Babcock & Wilcox), combustion air was preheated by extraction steam, and the heat from the combustion gases - no longer utilized in a boiler air preheater - was recovered in a low level economizer. Savings in equipment costs were claimed over a conventional cycle for the same overall efficiency. Emphasis was later shifted to combined cycle applications as in the Vitry-sur-Seine station, the extraction steam-preheated combustion air replaced by the oxygen-rich gas turbine exhaust, when the gas turbine is operating.

*See Appendix C.

In the U.S.S.R. extensive study of combustion air preheating by extraction steam has indicated that for plants
without boiler air heaters (the term boiler air heater referring to a regenerative flue-gas to combustion air heat
exchanger), the practice is thermodynamically justified
within certain limits, as it improves plant efficiency although accompanied by rising flue gas temperature to the
stack. For the practice to be effective for plants with
boiler air heaters, the ratio of water equivalent flows of
combustion air to flue gas has to be modified by splitting
the gas flow and introducing a low level economizer in
parallel with some of the intermediate feedwater heaters.
The rather complicated condensate and flue gas circuits
with this arrangement are shown in Figure 9.4. Air is
preheated by condensate to 160°F, as direct use of low
pressure extraction steam for preheating the air would
have resulted in large diameter steam lines and coils;
about 10% of the combustion gases and 22% of the condensate are routed through the low level economizer.
The study indicated that for new designs savings in fuel
of as much as 1.4% result, with the additional capital
investment recovered in 3 to 4 years. Modifying some
existing units in the 200 to 300 MW range for combustion
air preheating by extraction steam was also found feasible,
with resulting increase in the capacities of the units by
about 2% in the winter and from 0.4 to 1% in the summer,
without increasing the steam flow to the condenser.
Quick recovery of the investment in the additional equipment will also be realized.

In the United States combustion air preheating by extraction steam was investigated for the pioneer Eddystone
Station and stack gas heat was incorporated in the feedheating cycle between the last two steam extraction heaters. Part of the last point extraction steam is used to
preheat combustion air ahead of the regenerative air preheater to 112°F via an intermediate water loop. The
next-to-last extraction point was eliminated but a high
pressure feedwater heater was added, the upward dis-
placement of the extraction points resulting in increased
output and improved efficiency. The study, discussed in
one of the references, did not consider the more radical

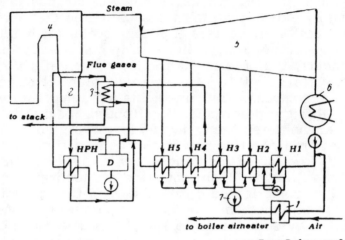

From Reference 9

System of regenerative air heating in combination with low pressure economiser.
HPH = high pressure heater; *D* = deaerator; *H*-1 ... *H*-5 = low pressure heaters; *1* – air preheating unit; *2* – boiler airheater; *3* – low pressure economiser; *4* – boiler; *5* – turbine; *6* – condenser; *7* – pump supplying water to air preheaters.

FIGURE 9.4

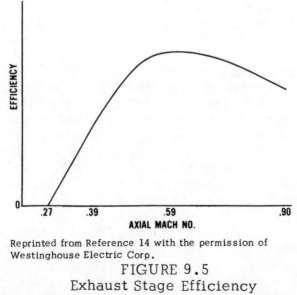

Reprinted from Reference 14 with the permission of Westinghouse Electric Corp.

FIGURE 9.5
Exhaust Stage Efficiency
vs. Axial Mach. No.

rearrangements of the extraction steam, flue gas and combustion air streams, reported as most promising by the foreign investigators.

Preheating combustion air by extraction steam to a boiler without a regenerative air heater has also been utilized in this country for simplified cycle steam plants intended for peaking and cycling service. Air is preheated by extraction steam to about 300°F, the gases leaving the economizer to the stack in the 500 to 600°F range. Good plant efficiency is reported, in spite of the high exit gas temperature, while substantial economics in equipment are realized by eliminating the air heater and from a reduced size turbine exhaust end because of the considerable reduction in the flow to the condenser that results from the steam extraction for air heating.

9.4   Exhaust End Selection

The high, widely varying turbine back pressures common with dry cooling systems complicate exhaust end size selection which, in the case of conventionally cooled plants, is a fairly straightforward choice based on fuel and hardware economics and cooling water temperatures.

For a typical single reheat, 2400 psig expansion line, steam specific volume varies in the ratios of 570 to 180 to 50 cu.ft./lb. as back pressure rises from 1 to 3-1/2 to 15 inches Hg. For a given exhaust annulus area steam velocity varies proportionately to volume flow fluctuations, and as the efficiency of the last turbine stage is a function of the steam velocity, Figure 9.5, for constant steam mass flow rate, efficiency is drastically reduced at high back pressures for a last stage designed to operate efficiently at low back pressures.

The rising back pressure has a twofold effect on turbine performance:
> Available energy in the steam to the turbine is reduced, the result of the lower cycle efficiency, as the expansion line and point (ELEP — the enthalpy of the steam leaving the last turbine stage) rises with the back pressure corresponding to the higher heat rejection temperature.

Internal efficiency of the last turbine stages - their
ability to convert the energy available in the steam
into work delivered to the turbine shaft - and leaving loss - the kinetic energy in the steam leaving
the turbine - are both strongly affected by back
pressure, their combined effect referred to as exhaust loss.

Exhaust loss shown for various exhaust annuli areas
in Figure 9.6 is a composite of "leaving loss", the kinetic energy of the exhaust steam, unrecoverable because
of the absence of any following turbine stages though
partially recoverable in diffusing type exhaust hoods,
which contributes the fairly straight line rise to the right
of the low points of the curves, and "turn up loss", the loss in
efficiency of the last turbine stages at low volume flows which
contributes the sharply rising segments to the left of the low
points of the curves; the small "hood loss", due to friction in
the turbine exhaust hood, is also included in the exhaust loss.

As the volume flow is reduced, reintrainment of the
steam discharged at the tip of the last stage blades takes
place near their base, as the last turbine stages reach
the point of actually absorbing rather than producing
energy in beating up and raising the temperature of the
recirculated steam. This accounts for the sharp rise in
exhaust loss at low volume flows. For a given turbine
exhaust and constant steam mass flow to the condenser,
rising back pressure results in reduced steam velocity
and a move to the left on the abscissa in Figure 9.6.

The combination of the "expansion line end point"
(ELEP) that reflects cycle or thermodynamic efficiency,
and the variation in exhaust loss which reflects the be-
havior of the exhaust end of the turbine at various volume
flows, results in the so-called "turbine end point" (TEP),
or "used energy end point" (UEEP) shown vs. backpressure
in Figure 9.7 for three exhaust end loadings.

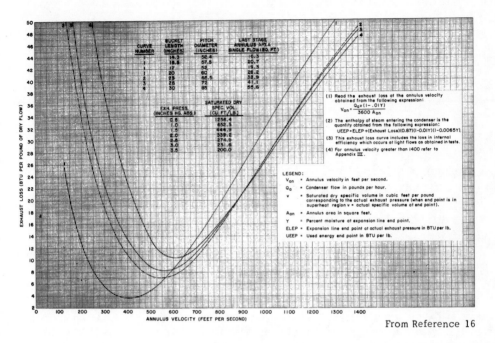

From Reference 16

FIGURE 9.6
Exhaust Loss as a Function of
Velocity in the Exhaust Annulus

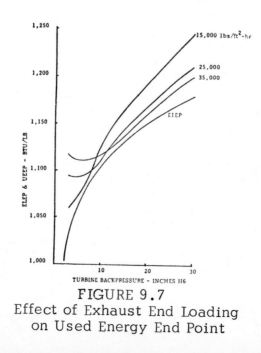

FIGURE 9.7
Effect of Exhaust End Loading
on Used Energy End Point

Exhaust end loading, expressed in pounds of steam flow per square foot of exhaust annulus area, or occasionally in kilowatts of turbo-generator output per square foot of exhaust annulus area times a factor for initial steam pressure,is a parameter widely used in engineering evaluation and selection of steam turbines.

At the usual low back pressures and large specific steam volumes encountered with once-through cooling systems, large and lightly loaded turbine exhaust ends provided for reasonable steam velocities and prevented choking* even in the winter months, allowing the turbine to utilize low ambients with correspondingly higher efficiencies. The additional cost of the larger, lightly loaded exhaust ends were justified on the basis of savings in fuel because of the low leaving loss and the better utilization of low ambients.

Domestic condensing turbine designs are currently limited by the manufacturers to exhaust end loadings of 15,000 lb/ft^2-hr, while lower limits were specified in the 1950's.

An upper limit is also imposed on back pressure, and operation with back pressure over 5 in. Hg is not recommended, or with exhaust steam temperatures exceeding 140°F, this temperature also imposing the limit for low load operation when the inefficiency of the last few stages causes exhaust steam temperature to rise and hood sprays have to be turned on. The high exhaust steam temperature also causes excessive differential expansion and distortion of the low pressure stages and casing, and sometimes the bearings have to be raised to allow proper clearances.

*Choking refers to the condition where steam velocity leaving the last turbine blade passages--designed as converging nozzles with varying degrees of reaction along their length--reaches sonic, and further decrease in back pressure with the resulting higher steam specific volume and higher velocity cannot be utilized for producing work but only increases leaving loss.

It was concern over low volume flows, high exhaust
end loadings, and the restrictions on operation and output
imposed by the higher back pressures with existing exhaust end load limits that caused the early negative attitude to dry cooling systems in this country. Indeed,
exhaust ends for higher back pressure turbines have to
be designed and supported to allow for the increased expansion, but the higher bending stresses in the shorter
blades of the smaller exhaust ends with the dry cooling
system present no serious problem in increasing exhaust
end loadings. The exhaust end load limit is dependent on
the aerodynamic and structural design of the last turbine
stages and, also, on the design of seals, bearings, and
casing supports. While domestic manufacturers currently
limit turbine exhaust pressures to 5 in. Hg, the corresponding limit for several large foreign manufacturers has
been 10 in. Hg, and often higher back pressures have been
allowed with slight modifications.

Domestic manufacturers have recently indicated that
they will offer turbines designed for the higher back
pressures encountered with air-cooled condenser power
plants when demand warrants their introduction, but
pricing attitudes differ from the foreign manufacturers.
Where foreign manufacturers offer the smaller, higher
loaded exhaust ends at lower prices that reflect the lower
material and fabrication costs, one domestic manufacturer has indicated that his prices for the higher loaded exhaust ends will not be lower but will be adjusted to ad-
here to the 15,000 lb/ft^2-hr exhaust end loading price.

The marketing attitude of domestic manufacturers to
tie turbogenerator pricing to output level and exhaust
end loading, rather than to the manufacturing cost of the
equipment, has another important determintal effect on
the design of air-cooled condenser power plants when providing for maintaining output at high ambients by increasing steam flow through the turbine, as discussed in the
next chapter.

As seen from Figure 9.7 high exhaust pressures favor
small exhaust ends to retain reasonable bucket-to-steam
velocity ratios at high backpressures and low volume
flows and to avoid the excessive turn-up losses that will
occur with large exhaust ends. The effect of exhaust end
size on capability and heat rate is incorporated in Figure 9.8;
i.e., with a heavily loaded exhaust end good performance
at low back pressures is traded off against limiting turn-
up loss at high back pressures, maintaining a fairly constant output over a wide range. Figure 9.9 illustrates this
point in more detail. As the exhaust end loading increases,
choking occurs at higher back pressures and the shaded
areas that indicate the operating range between choking
flow and high turn-up loss increase proportionally; the
ratio between the two back pressures at which the two
limiting phenomena occur is of the order of 1:4.

Of course, a number of selections is possible between
small and large exhaust ends, as shown in Figures 9.8 and 9.10,
the last for an 1100 MW, 1800 rpm nuclear steam turbine; and the choice of exhaust end also will reflect the
intended loading pattern, i.e., the hours of expected
operation at various ambient temperatures and the inci-
dence of a winter or a summer peak in the system.

It is of interest that part load performance of plants
equipped with dry cooling systems is excellent, efficiency remaining constant down to about one-third load.
As steam mass flow decreases, performance of the air-
cooled condenser improves because of the reduced thermal load, lowering back pressure and maintaining exhaust volume flow and turbine efficiency constant, rather
than introducing turn-up losses at the lower flows.

From a designer's point of view, the possibility also
exists of modifying exhaust end area by shutting off
steam supply to one or more low pressure sections of
a multiple exhaust turbine, these sections to continue rotating in near vacuum if on a common shaft with
powered sections; or by utilizing Baumann-type exhaust
ends where the last few turbine stages are formed to allow for side exhaust of part of the steam flow, and closing off the side exhaust area when the back pressure
rises, thus increasing steam flow in the axial direction.
The adoption of such methods is, of course, a matter of
incentives to overcome the higher turbine design and
manufacturing costs.

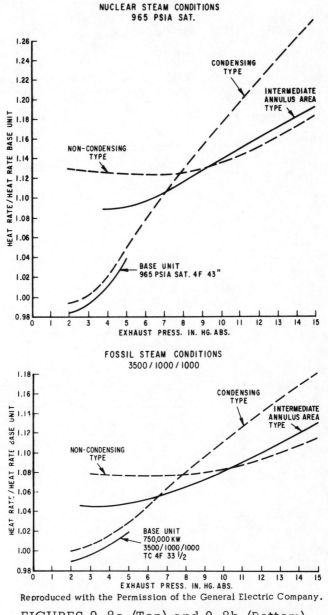

FIGURES 9.8a (Top) and 9.8b (Bottom)
Effect of Back Pressure on Turbine Heat Rate
for Various Exhaust End Areas as Estimated
by a Manufacturer for Nuclear and Fossil Steam Conditions

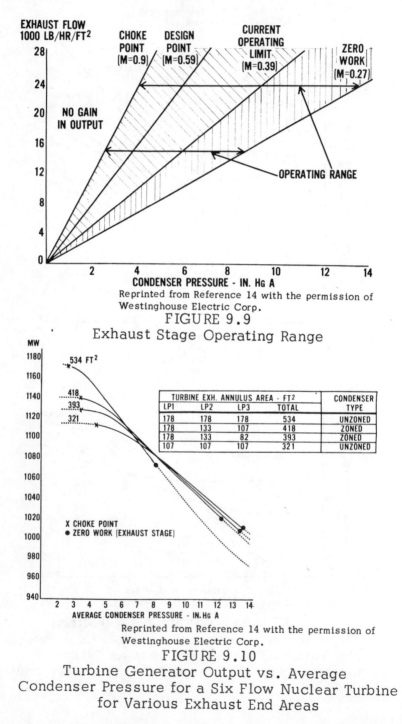

Reprinted from Reference 14 with the permission of
Westinghouse Electric Corp.

FIGURE 9.9

Exhaust Stage Operating Range

TURBINE EXH. ANNULUS AREA - FT²				CONDENSER TYPE
LP1	LP2	LP3	TOTAL	
178	178	178	534	UNZONED
178	133	107	418	ZONED
178	133	82	393	ZONED
107	107	107	321	UNZONED

X CHOKE POINT
● ZERO WORK (EXHAUST STAGE)

Reprinted from Reference 14 with the permission of
Westinghouse Electric Corp.

FIGURE 9.10

Turbine Generator Output vs. Average
Condenser Pressure for a Six Flow Nuclear Turbine
for Various Exhaust End Areas

REFERENCES

1. Bender, R.J., Supercritical boiler units, now in their third generation, gain tighter grip; Power, May 1970.

2. Petersen, H.J., The Economics of 2400 psig versus 3500 psig for Large Capacity Units; Proceedings of the American Power Conference, Volume XXV, 1963.

3. Hossli, W., Double Reheat; Brown Boveri Review, Vol. 53, No. 3, March 1966.

4. v. Flatt, F. and Scharffenberg, G., The Sonnefeld Process for Supercritical Thermal Power Stations (in German); E lektrizitaswirtschaft, Volume 67, 1968.

5. Leung P. and Moore R.E., Thermal Cycle Arrangements for Power Plants Employing Dry Cooling Towers; Transactions of the A.S.M.E., Journal of Engineering for Power, April 1971.

6. Cosar, P., The CESAS Cycle (in French); Revue Generale de Thermique, May 1967.

7. Cosar, P., Stevenin, M. and Widmer M., The CESAS Cycle for Heating the Air by Extraction Steam and the Water by the Gases (in French); VII World Power Conference, Moscow, 1968.

8. Senechaut, P., By Heating the Air with Extracted Steam and the Feedwater with FlueGas, Combined Cycle May Show Substantial Economy; Power, February 1968.

9. Apatovskii, L.E., Geltman, A.E. and Dmitrieva, N.T., Raising the Efficiency of Regenerative Air Preheating by Steam from Turbine Bleeds; Thermal Engineering (Teploenergetika), October 1967. HRVA Translations.

0. Ecabert, R. and Sirberring L., Optimizing the Stack Gas Temperature and Air Heating on Steam Generators Sulzer Technical Review, February 1969. Also in Combustion, April 1971.

11. Arnow, S.M., Incorporating the Heat from the Stack Gases into the Supercritical Turbine Heat Cycle at Eddystone Station; Proceedings of the American Power Conference, Volume 19, 1957. Also in Combustion, April 1957.

REFERENCES (Cont'd.)

12.   Clayton, W.H., Singer, J. & Tuppeny, W.H., Design
      for Peaking/Cycling, ASME Paper 70-Pwr-9. Also con-
      densed as "Design for Cyclic boiler operation; Power,
      February 1971.

13.   Silvestri, G.J. and Davids J., Effects of High Con-
      denser Pressure on Steam Turbine Design; Proceedings
      of the American Power Conference Volume 33, 1971.

14.   Oleson, K.A., Silvestri, G.J., Ivins, V.S. and Mitchell,
      S.W.W., Dry Cooling for Nuclear Power Plants; West-
      inghouse Power Generation Systems Report No. Gen-72-
      004, February 1972.

15.   Troyanovskii, B.M., et al, Designing Condensing
      Steam Turbines with Baumann Stages; Thermal Engineering
      (Teploenergetika) August 1967. HRVA Translations.

16.   Spencer, R. C., Cotton, K. C. and Cannon, C. N., A
      Method for Predicting the Performance of Steam Turbine
      Generators 16,500 kW and Larger; Transactions of the
      ASME, Journal of Engineering for Power, October 1963.

Chapter Ten

PLANT OPTIMIZATION AND EQUIPMENT SELECTION

10.1 Design for Minimum Present Worth Life-of-the Plant Cost

It is generally assumed that utility decisions relating to facilities construction and equipment selection, whether they are regulated investor-owned franchises or state agencies, are guided by the rule of supplying current and projected demand for electrical power, at a given level of reliability of service, for a minimum overall expenditure.*

In the final chapter, under "System Planning Considerations," there will be discussion of plant siting in relation to transmission line construction and of types of generating plant to meet varying duration loads in conjunction with efficiency and fuel costs. Also discussed will be the uncertainty imposed on plant costs and the heretofore single goal of minimum overall cost of power by the yet unformulated constraints of cost/benefit considerations that will be taking into account the growing environmental and aesthetic concerns.

In this chapter optimization of plant design for an air-cooled condenser plant will be considered, particularly as applied to equipment selection, in order to attain minimum cost of generation. (The related subject of thermal cycle considerations were discussed in the last chapter.)

In the United States system planning is usually carried out by the utilities, while plant design and equipment selection is undertaken by consulting engineering firms, with the exception of four or five very large utility systems that perform their own engineering through

---

*For suggestions of practices variant to this rule, see "Competition in the Nuclear Supply Industry"; A.D. Little, Inc. Report for the USAEC and the Dept. of Justice; NYO-3853-1; TID UC-2; December 1968. The ensuing denunciations appeared in "Utilities Get Dander up Over Criticism of Their Buying Practices"; Nuclear Industry, November-December 1968.

associated service companies. Some consultants also undertake the plant construction or construction supervision.

At this point, a brief digression becomes necessary to outline the method of evaluation of the alternative choices in power plant design. The life of a power plant is usually taken to be 30 to 40 years with little salvage value and a fixed charge rate is calculated that includes depreciation, Federal and State income taxes, insurance, local property taxes, and a regulated rate of return on the investment in the case of privately owned utilities. A plant loading schedule, i.e., the hours the plant is expected to operate at different loads during its lifetime and a projected fuel cost over the years is specified, the uncertainty inherent in future projections minimized by the decreasing use of the plant over the years and by the use of the interest rate to present worth, i.e., discount, future expenses. In the past depreciation was usually taken on a straight-line basis, but in recent years it has become increasingly important as its contribution is augmented by accelerated write-offs and investment credits intended to encourage new plant construction by all industries.

The Present Worth Life-of-the-Plant-Cost (PWLPC) is calculated on the basis of the Initial Investment (I), Fuel Cost (Fi) at some future year (i), Operating and Maintenance Costs (OMi) for the same year, Interest (r) and Fixed Change Rate (FCR) on the basis of the equation:

$$PWLPC = \sum_{i=1}^{n} \frac{I \times FCR + F_i + OM_i}{(1 + r)^i}$$

where (n) is the assumed life of the plant, in years. Of the various alternate plant designs considered, the lowest lifetime present worth cost design is selected, other factors, of course, being equal. Dividing the sum of the present worth costs by the total generation over the life of the plant yields the total present worth cost per unit of output, usually in mils per kW hr. Such analysis can be carried out in great detail by computers or by defining parameters such as levelized lifetime fuel cost, equivalent lifetime load factor, worth of a kW in incremental capability and auxiliary power, and worth of 1 BTU in heat rate improvement, for shortcut comparisons by the engineers of alternate equipment arrangements or machinery of different efficiencies.

Basically then, trade-offs involve an increased initial investment in equipment, in return for future savings in fuel because of improved efficiency or lower operating and maintenance costs. A brief description of the economic analysis is included in the easily obtainable first reference, while the second and third references treat the subject in more detail, with much valuable relevant information. Examples of optimization of specific systems and components are given in several other references and will not be elaborated upon here, as only factors peculiar to air-cooled power plant optimization will be considered.

Some information of a statistical nature but useful in recognizing the relative importance of the various power plant items has been included. Figure 10.1 illustrates the relative cost of the major power plant components, based on 1970 prices, together with cost projections to 1980. With costs varying widely from utility to utility and in a period of rapid escalation, Figures 10.2 and 10.3 - Average Investment per kW, 10.4 - Fuel Cost History, and 10.5 - Component Busbar Power Costs, are intended to convey only historical or representative cost information. In conjunction with Figure 10.1, examples of the transitory stage of the major power plant component costs are the cost of several planned systems of sulphur removal from the stack gases, estimated at between $15 to $35/kW, and deep water intakes or supplemental cooling for plants with once-through circulating water systems, at about $10/kW, not reflected in Figure 10.1. As design guidelines and technology have not matured in these areas, it is not known what these systems will contribute to overall plant costs. Air-cooled condenser system costs as reported by several published studies are shown in Table 14-II.

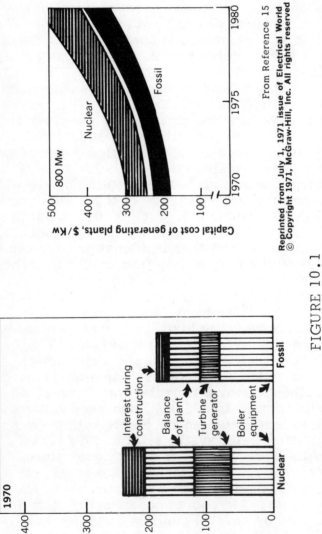

FIGURE 10.1

Capital Costs and Cost Trends of Generating Plants

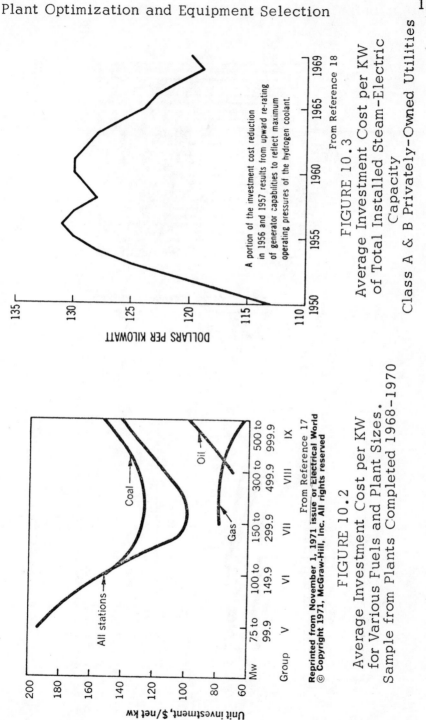

FIGURE 10.3
Average Investment Cost per KW
of Total Installed Steam-Electric
Capacity
Class A & B Privately-Owned Utilities

From Reference 18

A portion of the investment cost reduction in 1956 and 1957 results from upward re-rating of generator capabilities to reflect maximum operating pressures of the hydrogen coolant.

FIGURE 10.2
Average Investment Cost per KW
for Various Fuels and Plant Sizes.
Sample from Plants Completed 1968–1970

From Reference 17

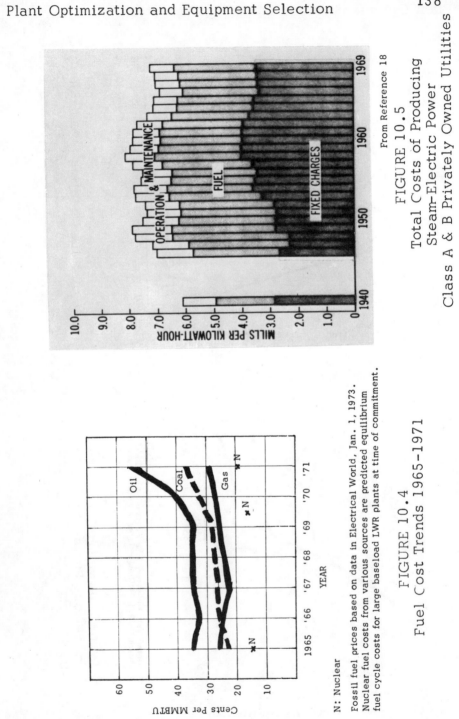

FIGURE 10.5
Total Costs of Producing
Steam-Electric Power
Class A & B Privately Owned Utilities

FIGURE 10.4
Fuel Cost Trends 1965–1971

N: Nuclear

Fossil fuel prices based on data in Electrical World, Jan. 1, 1973.
Nuclear fuel costs from various sources are predicted equilibrium
fuel cycle costs for large baseload LWR plants at time of commitment.

10.2 Factors Affecting the Design and Optimization of the Air-Cooled Power Plant

Since the boiler and turbogenerator have been the major cost items of a power plant installation, their selection has been the dominant consideration in power plant design, with the balance of plant optimized and designed on the basis of the boiler-turbogenerator selection. The inclusion of **air**-cooled condenser, expected to be a major cost item and a limiting factor in plant output, requires a reassessment of equipment selection guidelines and plant optimization procedures.

The first parameter to be considered in the design of an air-cooled condenser power plant for its drastic effect on output, cost and efficiency is the history of ambient air temperature at the site, usually summarized in a cumulative frequency distribution curve, of the form of Figure 10.6; indicating the hours that temperatures are exceeded for a typical year. Figure 10.7, indicative of the equally important daily fluctuation in temperature, is usually approximated by a sine curve, often of pronounced amplitude.

The need for the plant to maintain capability at high ambient temperatures requires careful selection and matching of the air-cooled condenser and turbine exhaust end and incorporation in the plant of special features for this purpose or, alternatively, the provision for other means of securing lost capability.

As usually carried out, air-cooled condenser optimization involves direct trade-offs of heat exchanger surface (and first cost of the air-cooled condenser installation) for initial terminal difference,which results in higher back pressure and lower efficiency. The optimum choice is affected by fuel economics, equipment costs, load factor, and the extent and importance of loss of capability at high ambients.

The subject of air-cooled condenser optimization and matching of the turbine exhaust end has been discussed in Chapters 4 and 9, and the selection of draft by fans or by natural draft tower in Chapter 6.

Part load performance and plant loading schedule are also important, as they enter into the determination of both the importance of lost capability and air-cooled condenser size selection. Given a projected plant loading

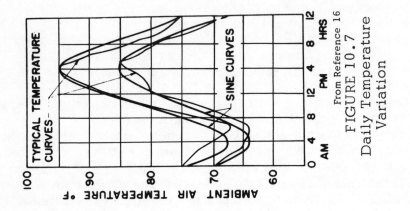

FIGURE 10.7
Daily Temperature
Variation

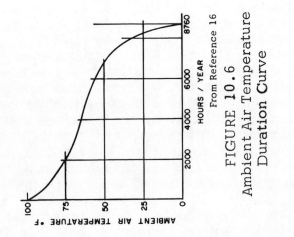

FIGURE 10.6
Ambient Air Temperature
Duration Curve

schedule and the temperature frequency and daily temperature variation curves, optimum air-cooled condenser
size and a matching turbine exhaust end can be selected on the basis of the heat rate vs. back pressure
curves of Section 9.4. An exhaust end can be chosen
to utilize high or low ambients or to better suit an average, little changing temperature, as the weather at the
site and the utility's load requirements dictate.

The following equations have appeared in Chapter 4,
where the symbols used were defined.

Heat exchanger surface for a given heat rejection load:

$$A_f = \frac{Q}{ITD} \left(\frac{1}{K_1 V_a \rho} + \frac{PH}{K_2 n V_w} + \frac{1}{K_3 n U_c} \right)$$

Part load performance of a given installation with a
fixed heat exchanger surface and varying heat rejection:

Mechanical Draft Installation   Natural Draft Tower
$$ITD = AQ^{.91}$$ $$ITD = AQ^{.75}$$

It should be noted that part load performance of air-
cooled condenser plants will actually be slightly better
than calculated from the performance of the air-cooled
condenser alone, as auxiliary power requirements can
be reduced at part load by reducing speed or shutting
off fans in mechanical draft plants, and, to a smaller
extent, by reducing circulating water power requirements in indirect mechanical and natural draft installations
to the point where savings in auxiliary power balance
loss from deteriorating air-cooled condenser performance.
Such optimization of operating procedures belongs properly in a later stage of power plant design where refinements in the plant design are considered, such as additional investment to power or control the fans or other
equipment traded off for part load economy or ease of
operation.

The optimization can be carried out by either of two
methods: With the first method the economic effect of
varying one parameter, the others remaining fixed at
reasonable values, is calculated, utilizing weighted
equivalent normalized life of the plant load factor,
fuel cost , etc., resulting in a series of graphs of the
effect of the air-cooling system parameters on cost, as

shown in Figures 10.8, 10.9, and 10.10. The other
method utilizes computer programs of varying complexity that successively calculate the present worth generation cost for successive changes in the variables
over predetermined ranges.

10.3 The Disputed Variables: Cost of Summer Replacement Capability and Features for Plant Capability
at High Ambients

Since mathematics and techniques of plant optimization
are fairly well agreed upon, it appears that there should
be fair agreement on selection of equipment and the cost
of generated power from air-cooled condenser plants.
That this is not the case can be explained by the fact
that two decisions, used as inputs in the optimization
procedure, are individual to the utility and the consulting engineer-designer of the plant, They are: the importance of loss of plant capability at high ambients together with the method of accounting for it by the utility,
and the plant designer's ability to minimize such loss by
judiciously oversizing other plant equipment or including
in the plant features for that purpose, all, of course, within the constraint of minimum overall cost of generated
power.
    The importance and method of accounting for loss of
capability varies from utility to utility. For a utility
with the maximum load occurring in the winter - a so-
called winter system peak - loss of capability in the
summer is obviously not very important and penalty for
such loss may even be neglected in comparative evalu-
ations of generating plant. For utilities with summer
system peaks, the importance of loss of capability of an
air-cooled condenser plant during the summer varies.
For a utility with large hydro or pumped storage capacity,
the loss is not as important as for an all-thermal plant
system.
    The usual methods of accounting for loss of capabil-
ity is by assigning a cost of replacement capability
penalty at or near the maximum yearly dry bulb temperature, in the range of $75-140/kW; by using an energy
replacement charge, where the cost of replacing the lost
generation throughout the year by less efficient or high

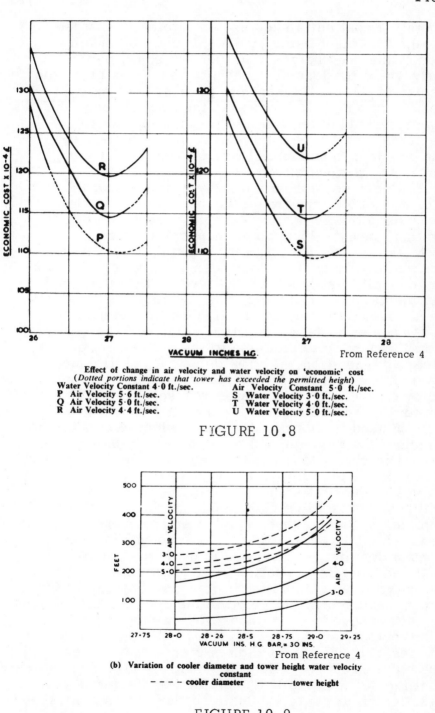

Effect of change in air velocity and water velocity on 'economic' cost
(*Dotted portions indicate that tower has exceeded the permitted height*)

Water Velocity Constant 4·0 ft./sec.	Air Velocity Constant 5·0 ft./sec.
P Air Velocity 5·6 ft./sec.	S Water Velocity 3·0 ft./sec.
Q Air Velocity 5·0 ft./sec.	T Water Velocity 4·0 ft./sec.
R Air Velocity 4·4 ft./sec.	U Water Velocity 5·0 ft./sec.

From Reference 4

FIGURE 10.8

From Reference 4

(b) Variation of cooler diameter and tower height water velocity
constant
– – – – cooler diameter ————tower height

FIGURE 10.9

operating cost units such as gas turbine is charged
against the plant being evaluated; or by a combination of the two charges. In the first case, particularly where the loss of capability penalty is high, the
design of the plant takes into account the maximum
yearly dry bulb temperature; in the second case, the
average dry bulb temperature is more appropriate.
Features that can be incorporated into the plant by
the designer to minimize loss of capability at high
ambients are several:

The obvious first option is by designing for the
higher dry bulb temperatures expected and effectively
oversizing the air-cooled condenser at the normal operating range, seldom an economical solution.

Methods used for obtaining peaking capability from
conventionally cooled plants can be used for maintaining capability at high ambients with air-cooled condenser plants. One method of fairly wide use with fossil-
fired baseload plants is shutting off the steam supply
to feedwater heaters, resulting in increased steam
flow through the turbine and to the condenser, and increased generation, at some decrease in efficiency;
an increase of 6 to 7% in output can usually be obtained by removing the top heater from service and about
3% for the second highest heater. A second method,
seldom used in this country but of wider use in Europe,
which offers the potential of considerably higher increase in output is bypassing high pressure steam from
the boiler to the turbine downstream of the admission
valves to the first stage of the high-pressure turbine,
which at the rated pressure limits steam flow to the
turbine and plant output. A third method accomplishes increased output by shutting off the steam supply to
combustion air preheating coils at a time coinciding
with reduced air preheating demand, the method being
effective when extensive air preheating by extraction
steam has been incorporated in the cycle by the designer. The above three methods of obtaining increased
plant output, or, conversely, of nullifying the effect
of loss of capability at high ambients, have in common
the requirement for a flexible exhaust end loading policy
that runs contrary to U.S. turbine manufacturers pricing
practices relative to exhaust end flow limits discussed

in Section 9.4, which will become accentuated at the
higher exhaust end loadings for air-cooled condenser
plants. For all three methods to be economically viable,
they require to a varying degree that the additional steam
flow through the turbine exhaust end at times of high
ambient not penalize exhaust end selection by dictating,
because of flow limits set by pricing considerations, the
selection of a larger exhaust end than required at the normal operating range. It is of considerable interest that
foreign turbine manufacturers are quite accommodating in
this respect. The ability of the turbine exhaust end to
operate adequately in the anticipated operating range,
with the back pressure further augmented as a result of
increased steam flow is also essential.

The three methods also have in common the need for
increased output from the boiler at such times, within
the specified design and temperature limits, as in every
case overfiring the boiler will be required to make up for
the heat input from the diverted steam flows to the feedwater (first method), to combustion air (third method), or
for the increased steam flow (second method).

Boiler manufacturers have indicated that such increased output capability for specified periods of operation can be built into current boiler designs for
limited increments in output at an incremental output
cost of 15-25% of the base output cost, or that designs
can evolve incorporating novel features that can result
in considerable increase in output over the base, at
varying loss in base and incremental boiler efficiency.
Auxiliary boilers can also be incorporated in the plant
design for the purpose of supplying the additional required steam output at such times.

Comparison of the treatment of loss of capability
in the studies summarized in Table 14-II is of interest.
The studies originating in the United States assume cost
of replacement capability based on gas turbines. The
choice is questionable, since gas turbines also suffer
loss of capability at high ambients and oversizing the
replacement power source must be resorted to, resulting in considerable penalty. In two of the studies,
the gas turbine replacement capability amounts to one-
third of the additional equipment cost of the air-cooled
condenser plant. The foreign studies either do not include a loss of capability penalty or include features
of alleviating such loss of capability in the cost of

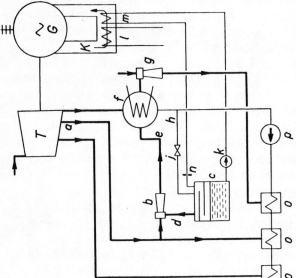

Specific arrangement of supplementary cooling with steam ejector showing the insertion of the installation in the thermal system of the generating station.

G, generator; K, circulating air; T, steam turbine; a, bleed point; b, steam ejector; c, evaporator; d, suction line of the steam ejector; e, counterflow line of the steam ejector; f, steam turbine condenser; g, vacuum pump; h, condensed water line; j, regulating valve; k, circulating pump; l, cooling system; m, supplementary cooling system; n, reducing disc; o, feed water heater; p, turbine condensate pump.

From Reference 14

FIGURE 10.11

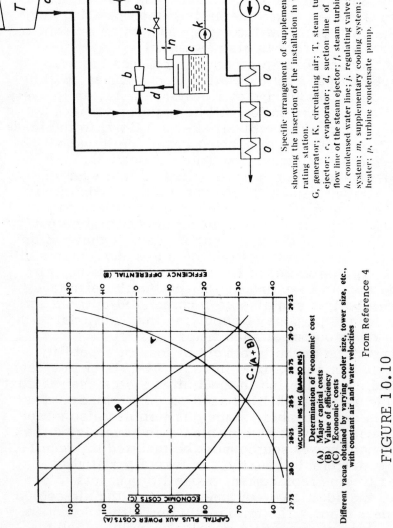

(A) Major capital costs
(B) Value of efficiency
(C) 'Economic' costs

Determination of 'economic' cost

Different vacua obtained by varying cooler size, tower size, etc., with constant air and water velocities

From Reference 4

FIGURE 10.10

the plant. It is also of interest that of the large air-cooled condenser plants built (aside from Rugeley, built with an oversized air-cooled condenser, for which the question of loss of capability at high ambients has not been raised), Ibbenburen has been designed to maintain capability to 95°F and Utrillas to 86°F.

Other means of maintaining capability at high ambients that belong in a different category are by augmenting heat transfer with a fine spray of water on the exchanger fins and tubes, or by precooling the air by evaporative cooling at such times, similar to the process industry practice of supplementary spray or water trim-cooling at high ambients. The acceptability and extent of the practice depends, of course, on the availability of the required amounts of water for collection and intermittent use. In mechanical draft installations, the selection of multispeed fans offers the option of operating the fans at the higher speed to improve heat rejection at high ambients, while methods of augmenting draft in natural draft towers at high ambients have also been suggested.

Finally the problem of total water unavailability, even for a small evaporative tower to serve the auxiliaries' cooling needs such as generator hydrogen and various bearing oil coolers, has resulted in an ingenious solution by Heller and associates in serving such cooling loads. In the case where the acceptable operating temperature of the primary fluids cooled (i.e., hydrogen, bearing oil, etc.) cannot be reached, even when using water from an air-cooled tower of a closer approach than the main air-cooled condenser, the incorporation of a simple refrigeration plant utilizing a steam air ejector in the power plant's cycle will serve such loads in a simple and reliable manner. To reduce additional equipment, the ejector can discharge the steam and reject heat in the power plant's jet condenser, as shown in the scheme in Figure 10.11, or to a separate air-cooled condenser.

REFERENCES

1. A Survey of Alternate Methods for Cooling Condenser Discharge Water, Operating Characteristics and Design Criteria; Environmental Protection Agency, Water Quality Office. Water Pollution Control Research Series Report 16130 DHS 08/70.

2. Bartlett, R.L., Steam Turbine Performance and Economics; McGraw Hill, Inc., 1958, Chapters 9 through 12.

3. Bary, C.W., Operating Economics of Electric Utilities; Columbia University Press, 1963.

4. Daltry, J.H. and Cheshire, L.J., A Closed Circuit Cooling System for Steam Generating Plant; The South African Mechanical Engineer, February 1960.

5. Daltry, J.H., Discussion to "Rugeley Dry Cooling Tower System" by Christopher, P.J. and Forster, V.T.; Proceedings of the Institution of Mechanical Engineers, Volume 184, Part 1, No. 11.

6. Smith, E.C. and Larinoff, M.W., Power Plant Siting, Performance and Economics with Dry Cooling Tower Systems; Proceedings of the American Power Conference, Volume 32, 1970.

7. Research on Dry Type Cooling Towers for Thermal Electric Generation, Part I; Environmental Protection Agency, Water Quality Office. Water Pollution Control Research Series Report 16130 EES 11/70.

8. Seippel, C., Steam Turbines for Variable Load and Overload Operation (in French); World Power Conference, Lausanne 1964, Paper No. 104. Also: Hossli, W., Overloading a Steam Plant to Cover Peak Loads; Brown Boveri Review, Volume 53, No. 3, March 1966.

9. Hawley, C.F., Peaking Capacity in Steam Generating Units; Power Engineering, July 1962.

10. Schoonman, W., Air Cooled Steam Condensers for Power Plants; Paper presented to the (Australian) Society of Mechanical Engineers Symposium on Decentralization of Energy Production in Relation to Available Water Resources, Sydney, Australia, Sept. 1970.

11. Miliaras, E.S., Low Cost Capacity for Peaking Generation; Transactions of the A.S.M.E., Journal of Engineering for Power, January 1968.

REFERENCES (Cont'd.)

2. Reducing Costs of Dry Cooling Towers for Electric Power Plants; Power Systems Engineering Group Report No. 13. Mass.Institute of Technology, May 1969.

3. Mathews, R.T., Some Air Cooling Design Considerations; Proceedings of the American Power Conference, Volume 32, 1970.

4. Mandi, A., Urbanek, J., and Heller, L., Tests Aimed at Improving the Cooling of Turbo-alternators (Part II); GIGRE 1956, Paper 132.

5. Plant Capital Costs Spiraling Upward; Electrical World, July 1971.

6. Reti, G. R., Dry Cooling Towers; Proceedings of the American Power Conference, Volume XXV, 1963.

7. 17th Steam Station Cost Survey; Electrical World, November 1, 1971.

8. 1970 National Power Survey; The Federal Power Commission.

Chapter Eleven

POWER PLANTS FOR PEAKING/CYCLING WITH AIR-COOLED CONDENSING SYSTEMS

11.1 Power Plants for Limited Hours of Operation

The varying demand for electric power is illustrated in Figures 11.1a and 11.1b. Figure 11.1a is a typical utility system daily load curve while Figure 11.1b is a load duration curve, i.e., the total number of hours per year at different system loads. To economically supply the daily and seasonal demand for electric power, utilities rely on several types of generating plants.

For a limited number of hours of operation a year, the so-called low load factor range, high fuel cost or poor efficiency plant will be acceptable if it is counterbalanced by low equipment cost and other factors are favorable, such as quick starting for availability on short notice and ease of siting and operation. An outstanding example is the gas turbine, installed by many utilities in increasing numbers and for a larger percentage of the generating mix, where the high fuel cost and poor efficiency are acceptable because of the lower first cost, quick starting, ease of siting and short

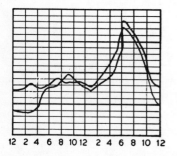

FIGURE 11.1a
Daily Load Curves

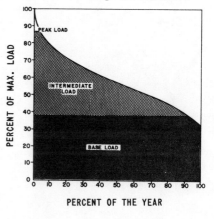

PERCENT OF THE YEAR

Load Duration Curve

delivery schedules. Another example is the fossil-
fired peaking/cycling steam plants being installed
by utilities to replace older, obsolescent equipment,
that was traditionally operated to cover system peaks.
These plants, the first of which went into operation
in 1968 with about a dozen more planned by utilities
in ratings up to 400 MW, have shown excellent operating economies and demonstrated the soundness of
the concept of the lower efficiency and cost steam
plant for low load factor operation. (The plants were
first proposed and marketed unsuccessfully in 1963.)
    The peaking/cycling steam plants are built
around a low-cost oil or gas fired boiler at conservative steam conditions, usually 1800 psig pressure and
950°F superheat temperature with reheat to 950°F, and
are designed expressly for cyclic service. The plants
are provided with only three or four feedwater heaters
and a steam turbine capable of cyclic operation, while
steam air heating is usually utilized for the combustion air instead of air heating by the flue gases. Quick
starting features and simplified operating procedures
are also incorporated in the design.

11.2 Advantages of Air-Cooled Steam Plants for
      Cycling Service

The advantages of power plants with air-cooled condensing systems are often stated as freedom from the siting
requirement near cooling water supplies and transmission
line construction savings by locating near load centers,
balanced against higher fuel transportation costs and
lower plant efficiency. The peaking/cycling steam
plant is an area of power plant design where these advantages offered by the air-cooled plants can be utilized and the disadvantages reduced.
    For example, a plant intended to operate for a six-
hour weekday period will total about 1500 hours of operation per year, or about a 17% plant capacity factor, and
will require several times less fuel than a base load
plant, proportionally reducing fuel transportation costs.
Similarly, the lower plant efficiency resulting from the
higher back pressure of the air-cooled condenser will

not be a great disadvantage with the low plant capacity
factor. Indeed, the back pressure will optimize at a
higher value, proportionally reducing air-cooled condenser cost; and by locating near load centers, transmission
line construction is reduced and the sites near cooling
water sources can be better utilized for base load plants.

A peaking/cycling steam plant with air-cooled
condenser will probably utilize more feedwater heaters
and higher steam conditions than its conventionally
cooled counterpart. As discussed in the chapter, "Thermal Cycle Arrangements," extensive preheating of the
combustion air by steam extraction instead of a regenerative air heater will be employed, a practice already
utilized by peaking/cycling steam plants. Quick starting will be a major requirement for such plants and the
air-cooled condenser will have to be designed to maintain the exceptionally fast starting capability realized
by some peaking/cycling steam plants-less than 20
minutes to full load from a cold start in the case of the
150 MW foreign installation described in the fourth
reference. The Heller system plant can be designed
for such quick start-ups, according to the manufacturers, while the direct system is reportedly capable
of maintaining vacuum for long periods while on stand-
by without freezing. Another proposed design for a
cycling operation air condenser plant utilizes a hybrid
Heller system, with a surface condenser instead of the
jet condenser and with an antifreeze solution in the
thus separated circulating water circuit that comprises
the surface condenser, circulating water piping and
air-cooled heat exchanger elements. With such plants
the loss of capability at higher ambients will not be
severe, as the highly loaded small turbine exhaust
ends that will be selected because of the higher back
pressures that the air cooled system will optimize for
low load factor operation, are less susceptible to loss
of capability with rising back pressure at higher ambients. But it is possible at moderate cost to incorporate in the plant design features for maintaining output even at extreme high ambients. Additional boiler
capability can be provided with overload bypass valves
past the first few high pressure turbine stages, or the

steam for preheating the combustion air can be diverted to the steam turbine with provisions for overfiring
the boiler. The additional steam flow through the turbine can be used, at a loss in plant efficiency, to
maintain capability at extreme high ambients.

11.3 Relative Merits of Air-Cooled Steam Plants and
      Gas Turbines for Peaking/Cycling Loads

The fossil fired peaking/cycling steam plant, when
equipped with an air-cooled condensing system, offers excellent prospects for supplying low cost peak-
ing/cycling power for sites removed from cooling
water or near load centers where the gas turbine has
been the system planner's only choice.
      A comparison of the relative merits and disad-
vantages of gas turbines and air-cooled condenser
plants for peaking/cycling service will disclose that
the gas turbine is certainly the undisputed choice for
emergency, spinning reserve and black-start applications and for peaking service for operation up to perhaps 1,500 hours per year. But for mid-range appli-
cations-1,500 to 3,500 hours of operation per year-
the higher first cost air-cooled condenser power plant
offers the same freedom of siting as the gas turbine,
at a much better heat rate and with a low fuel cost.
      Currently marketed large industrial gas turbines
have heat rates of 14,000 to 15,000 BTU/kwhr, based
on the Higher Heating Value of the fuel, as compared
to between 10,000 to 12,000 BW/kwhr for the peaking/
cycling steam plant with air-cooled condenser. Gas
turbines are also limited to gaseous fuels and the
lighter distillate oils, while the peaking/cycling steam
plant can be fired with residual fuel that has traditionally enjoyed a price advantage over the light distillates.
The combined effect of better heat rate and lower cost
fuel for the peaking/cycling steam plant results in a
fuel cost component about half that for the gas turbine.
Experience in utility systems with a large number of
gas turbines indicates that maintenance costs per kwhr
generated are about three times that of a steam plant.
The high fuel and maintenance costs of gas turbine gen-

erated power become increasingly significant as the
hours of operation are extended; above 1,500 hours
of operation per year the advantage of the low first
cost of the gas turbine is lost to the higher first cost
but lower fuel and maintenance cost of a peaking/
cycling steam plant with air-cooled condenser.

Tables 11-I and 11-II, compare representative
values and total generation costs for the gas turbine
and two peaking/cycling air-cooled steam plants for
six-hour weekday operation or about a 17% plant
capacity factor. The costs are based on estimates
for 1972 initial operation; the effect of increasing
hours of operation is illustrated in Figure 11.2.

Fuel price is an essential parameter in power
plant selection, increasing in importance as the
projected plant capacity factor rises. Customarily
there has been a price differential between the residuals fired in power boilers and the lighter distillates
burned in gas turbines, the lighter distillates priced
at between 50% and 100% higher. Present shortage
of low sulphur residuals has periodically narrowed
the gap, but there is indication that when the present
fuel market stabilizes, the former pricing differential will return. "Low Cost Fuels Coming for Gas
Turbines" is a recurrent heading in trade magazine
articles of the last several years, but such a satisfactory middle distillate or blended fuel has not yet
been proved.

A comparative examination of the loss of capability at high ambients of the two types of plant
is of interest. Present industrial gas turbine designs
will suffer a 15% loss of their 60°F capability* at

*The loss of capability for the gas turbine is deter-
mined from output correction factors vs. ambient
supplied by manufacturers. For the loss of capability of the air cooled steam plant see Figure 9.8b
Chapter IX, the "Intermediate Annulus Area" curve
at exhaust pressures of 6" Hg (60°F) and 15" Hg
(100°F), an ITD of 80°F is assumed.

TABLE 11-I
Plant First Costs and Other Data

	Gas Turbine	Air Cooled Steam Plant "A"	Air Cooled Steam Plant "B"
Installed cost, $/kw	90	170	140
Heat rate, BTU/kwhr	14,000	10,000	12,000
Fuel cost, c/MMBTU	90	60	60
Fixed charge rate	15	15	15
Hours of operation	1,500	1,500	1,500

TABLE 11-II
Comparison of Total Generation Costs

	Gas Turbine	Air Cooled Steam Plant "A"	Air Cooled Steam Plant "B"
Annual plant cost, mils/kwhr	9.0	17.0	14.0
Fuel cost mils/kwhr	12.6	6.0	6.0
Oper. & maint. cost mils/kwhr	2.5	1.0	1.0
Total Generation Cost	24.1	24.0	21.0

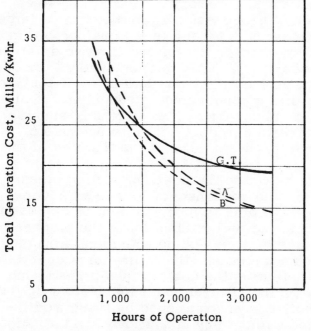

FIGURE 11.2
Comparison of Generation Costs by Gas Turbine
and Air-Cooled Steam Plants

100°F ambient. The only means of reducing gas turbine
capability loss at high ambients is by water spray cooling of the incoming air, raising the original question of
water availability, while operating the gas turbine at
above base rating capability for the same purpose,
sharply increases maintenance costs and is to be avoided.
The air-cooled steam plant, without the features for
sustaining capability at high ambients discussed in
the previous section, will only suffer an 8% loss of
capability over the same temperature range.

No discussion of plants for mid-range generation
is, of course, complete without mention of the combined
cycle plants, that, after being offered unsuccessfully
for several years, are finally being ordered by the utilities. But combined cycle plants only partially reduce and do not eliminate the need for circulating cooling water, or evaporative makeup and, as a result,
will not compete with the air-cooled cycling steam
plant for the applications suggested here. In this
context the concept of a combined gas turbine-steam
turbine cycle plant unsuccessfully marketed by a large
consulting engineering company is of considerable interest and worthy of note. Described in some detail in
the last reference, the 200 MW pre-engineered, modular design plant derived three quarters of its output from
an air-cooled, direct condensing, steam plant. Quick
startup, black start capability, an extremely simple
steam cycle but with a modest heat rate because of
waste heat recovery, were some of the features of the
plant.

Another area where the peaking/cycling steam
plant with air-cooled condenser may have an advantage over the gas turbine is in the amount of oxides of nitrogen emitted during fuel combustion. The proposed
methods of reducing nitrous oxide emissions from gas
turbines by steam injection will require large quanti-
ties of water, defeating the gas turbines prime siting
advantage. Reducing nitrogen oxides emissions in the
steam plant's boiler is accomplished by firing modifications only and without any consumptive water usage,
leaving the peaking/cycling steam plant with air-cooled
condenser the only plant truly independent of a source
of water.

REFERENCES

1. Heyburn, D. E. and Brandon, W. W., Application of
    Low Cost Thermal Power for Peaking; ASME Paper 64-
    PWR-12. Also: Ready-Reserve Power Plant; Mechan-
    ical Engineering, January 1964.

2. Clayton, W. H., Singer, J. G. and Tuppeny, W. H.,
    Jr., Design for Peaking/Cycling; ASME Paper 70-PWR-
    9. Also: Designing for cyclic boiler operation; Power,
    February 1971.

3. Werner, R. P. and Singer J. G., 400 MW Cycling Unit
    Slated for June 1971; Electrical World, August 19, 1968.

4. Schenkenberg, H., A Cycling/Peaking, Fast Starting
    Steam Plant; Combustion, March 1971.

5. Gilliom, J. R., Operating Experience with Gas Turbine
    Peaking Units on the Commonwealth Edison System;
    Proceedings of the American Power Conference, Vol. 33,
    1971.

6. Dibelius, N. R., Hilt, M. B. and Johnson, R. H., Re-
    duction of Nitrogen Oxides from Gas Turbines by Steam
    Injection; ASME Paper 71-GT-58.

7. Bagwell, F. A., et al, Oxides of Nitrogen Emission Re-
    duction Program for Oil-and Gas-Fired Utility Boilers;
    Proceedings of the American Power Conference, Vol. 32,
    1970.

8. Miliaras, E. S., Discussion of "Thermal Cycle Arrange-
    ments for Power Plants Employing Dry Cooling Towers"
    by Leung, P. and Moore, R. E.; Transactions of the ASME,
    Journal of Engineering for Power, April 1971 and January
    1972.

9. Combined-Cycle Plant Uses Dry Cooling - 200 MW
    Combined Cycle Package Offered; Electrical World,
    December 1, 1970.

Chapter Twelve

NUCLEAR POWER PLANTS REJECTING HEAT TO AIR

12.1 Air Cooling for Light Water Reactor Plants

In the United States, air-cooled condensers for nuclear
power plants have been looked upon with favor and considerable AEC interest; several studies have reported
the effects of air cooling on nuclear generation economics.*
     Air cooling for nuclear plants removes the last of
the siting constraints, permitting the location of such
plants near load centers as far as current safety restrictions relating to population density allow. The other
siting constraint that applies to fossil fired plants, i.e.,
plant location near a source of fuel or fuel transport terminal, does not apply to nuclear plants as the transportation cost component of nuclear fuel is small and nearly
independent of plant location.
     Other reasons for the favorable attitude towards air
cooling for nuclear power plants include the assumptions
that the high equipment cost of a nuclear plant can
easier absorb the additional cost of the air-cooling
equipment and that the low nuclear fuel cost justifying
high equipment costs will also accommodate the loss in
efficiency resulting from the higher back pressure of an
air-cooled condenser.
     But several pronounced drawbacks of air-cooled con-
densers for the currently marketed Light Water Reactor**
plants should be given consideration:
          They require about 50 percent more air-cooled
          heat exchanger surface per kW output, reflect-
          ed in air-cooled condenser cost and fan power
          requirements or cost of natural draft tower struc-
          ture, compared to current fossil fired plant de-
          signs.

* See Table 14-II.
**The term encompasses Pressurized Water Reactor
  (PWR) and Boiling Water Reactor (BWR) plants.

There is much faster loss of capability and deterioration in performance at high ambients,
again compared to current fossil plant designs.
The performance of the lightly loaded, high
inlet moisture, multi-flow exhaust ends of
the nuclear steam turbines deteriorates about
twice as fast for a given rise in back pressure
than that of a modern, high-pressure reheat
steam turbine. This comparison applies to
current exhaust end loadings for the respective designs.

The projected load factor for a nuclear plant
is higher compared to a fossil plant, as the
former is intended--and its high capital cost
is justified--on the basis of baseload operation, while fossil plants will be assigned to
the intermediate load range. As a result, operating penalties for the nuclear plants are
assessed over a greater number of operating
hours.

Safety considerations may dictate against the
Heller System's shared condensate-circulating water circuit. A conventional surface condenser can replace the jet condenser in this
case with some loss in thermodynamic performance and greatly simplified operation. This
hybrid Heller System is currently proposed by
a large U.S. manufacturer for light water
reactor plants.

The sheer size of an air-cooled condenser installation for a nuclear plant - 10 to 15 acres
for a 1000 MW plant,as nuclear plants are not
popular or economical in the smaller ratings -
makes the decision to go air cooled with the
next nuclear plant a very difficult one for a
utility. As air-cooled condensers are considered an emerging technology in the United
States, it will be difficult to make the first,
giant step in the new direction. The decision
to go to air cooling should be much easier for
a smaller rating, intermittently operated and
less sensitive fossil fired plant.

Incidentally, there should be no lack of space or a pre-
mium for the required acreage for the air-cooled condenser for a nuclear power plant as regulations regarding an
exclusion area for such plants should provide adequate
space for air-cooled condenser installations.

12.2 Reduced Heat Rejection with the Steam Cycle
HTGR Plants

The large heat rejection requirements of the light water
reactor plants, as compared to fossil fired plants or to
other nuclear cycles, becomes an important factor when,
as a result of environmental restrictions on thermal discharges, about half of the nuclear plants on order or under construction will utilize wet cooling towers in open
or closed systems. The water consumption of such
plants will be impressive, a 1000 MW LWR plant cooled
by an evaporative cooling tower consuming for evaporation and blowdown about 20 MGD - adequate water supply for a city of three hundred thousand.

The third commercial U. S. nuclear plant design, the
High-Temperature Gas-Cooled Reactor plant, rejects
only about two-thirds the waste heat of a LWR plant of
the same electrical output. The HTGR plant utilizes a
pyrolitic carbon-coated fuel cooled by helium heated to
high temperature, which in turn produces high pressure
and temperature steam for a modern practice Rankine
cycle steam plant. The LWR plants utilizing metal clad
fuel are restricted to lower fuel element temperatures
producing in turn lower quality steam for a less efficient
cycle.*

It is worth noting that reduced heat rejection by this
reactor type plant has been mentioned as a factor in the
choice of the HTGR plants recently ordered, which will
utilize cooling towers.

*The HTGR operates on the uranium-thorium fuel cycle
with a uranium-thorium carbide fuel element and a gra-
phite-moderated helium-cooled core. The PWR and BWR
have uranium oxide fuel, clad in stainless steel or zir-
conium, in a water-moderated, water-cooled core. Until recently, the 40 MW Peach Bottom Plant, about to be
retired, and the 330 MW Fort St. Vrain Plant, which was
to be completed in 1972, were the only two HTGR plants
ordered.

The following table summarizes the representative characteristics of the Light Water Reactor Plants,
High-Temperature Gas-Cooled Reactor plants, and the
currently built efficient fossil fired plants; the figures
are for 2" Hg back pressure, usual with 60°F water open
cycle cooling.

TABLE 12.1
EFFICIENCY AND HEAT REJECTION BY LARGE POWER PLANTS

	LWR Plants.	HTGR Plants	Fossil Fired Plants
Steam Conditions	950 psig, sat. with 2 reheats	2400 psig 950°F/1000°F	3500 psig 1000°F/1000°F
Efficiency	33.8	39.1	38.5
Net Plant Heat Rate	10,100	8,750	8,860
Heat Rejected* per kw output	6,690	5,340	4,470

In the case of the fossil fired plant, an additional 980
Btu per kw generated will be discharged to the atmosphere through the boiler stack (at a boiler efficiency of
90%).

12.3 The HTGR with a Direct Gas Turbine Cycle.

An intriguing prospect has been raised for the High-
Temperature Gas-Cooled Reactor with a helium gas
turbine cycle for power generation. In the currently
available HTGR designs helium acting as core coolant

*About three percent of the heat does not leave the cycle
through the main cooling system but is dissipated through
generator losses and at the bearings and the motor drives
of the various plant auxiliaries.

and heat transfer medium transfers the reactor-generated heat to a steam generator for a conventional Rankine power cycle. In contrast to the Light Water Reactors generating low temperature saturated steam, the HTGR pyrolitic carbon fuel coating allows high helium temperatures to the steam generator and a Rankine cycle of the same high temperature and efficiency as modern fossil fired steam plants. In a proposed novel cycle configuration for the HTGR Plant, shown in Figures 12.1 and 12.2, the Rankine cycle is dispensed with and water is no longer the working fluid in the power generating cycle. Instead, the helium leaving the reactor at about 1500°F and 1000 psig is expanded through a turbine to about 400 psig to generate power and drive the compressors for recompressing the helium, after heat rejection to the air between 340°F to 100°F through an intermediate water loop.

The intermediate heat rejection water loop is required as a barrier to sudden depressurization and loss of the high-pressure helium coolant in the event of fracture of the helium mains; the reactor plant will be of the so-called integrated design, where all the helium loop components will be housed in a prestressed concrete pressure vessel.

The closed gas turbine cycle for this application, the so-called A-K Cycle, has been pioneered in Germany with several small installations that have been in operation since the mid fifties, having air as the working fluid. A 25 MW helium turbine with a nuclear reactor heat source and conventional cooling was to be completed in Geesthacht, Germany, in 1972; and a 600 MW helium turbine, also with a nuclear reactor plant, is discussed for 1976 completion. Several advantages are inherent in the cycle.

Reduction of the size of turbomachinery, as the high velocity of sound in helium allows high flow velocities and smaller cross sections in the turbomachine.

Reduction of heat exchanger surfaces resulting from the excellent heat transfer properties of helium.

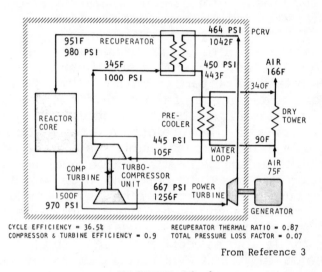

CYCLE EFFICIENCY = 36.5%   RECUPERATOR THERMAL RATIO = 0.87
COMPRESSOR & TURBINE EFFICIENCY = 0.9   TOTAL PRESSURE LOSS FACTOR = 0.07

From Reference 3

FIGURE 12.1
Air-Cooled HTGR with Helium Gas Turbine Cycle

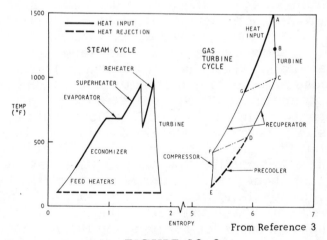

From Reference 3

FIGURE 12.2
T-S Diagrams for an HTGR-Steam Plant and the
Air-Cooled HTGR-Helium Gas Turbine Plant

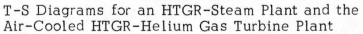

Good efficiency throughout the plant load range
by varying the amount and pressure levels of
the working fluid in the circuit while maintaining pressure ratios and constant near-optimum
flow velocities through the turbine.

Other advantages of the cycle in nuclear applications
result from the inert properties of helium and fission
products retaining properties of the carbide fuel. But
the important advantage of the cycle for air cooling
results from its suitability for rejecting heat to air.
In the Rankine cycle heat rejection takes place nearly
isothermally as steam condenses. As a result, if air
is used as the ultimate heat rejection medium, its
temperature rise has to be kept low to prevent high
back pressure in the steam turbine, since the Rankine
cycle efficiency is very sensitive to turbine back pressure. The multipressure condenser is an effort to cir-
cumvent the isothermal condensation of the Rankine
cycle plants and reject heat reversibly over a temperature range. The requirement for low air rise when
Rankine cycle plants are rejecting heat to the air results in the need to handle large air flows at very low
pressure drops and the consequent need for several
large natural draft cooling towers or a great number of
fans to create the necessary air flow. In contrast the
working fluid in the gas turbine cycle takes place at
constant pressure over a temperature range and does
not undergo a phase change during heat rejection. In
the proposed air-cooled, helium turbine-HTGR Plant,
heat rejection to the air will be between 100° and 340°F.
The proposed helium turbine-HTGR Plant, with a power
cycle as in Figures 12.1 and 12.2 and air cooling, will
have about the same efficiency as a HTGR with a Rankine
cycle and a wet cooling tower. To compensate for this
high average heat rejection temperature and the cycle
inefficiency resulting from the large work input to compress the helium, as compared to the Rankine cycle
where the work input for compressing the working fluid
in the liquid state is small, the top cycle temperature
in the helium cycle will be about 1500°F*, compared
to 1000°F for the Rankine cycle.

*This is conservative gas turbine inlet temperature,
present open cycle combustion gas turbines operating
with considerably higher inlet temperatures.

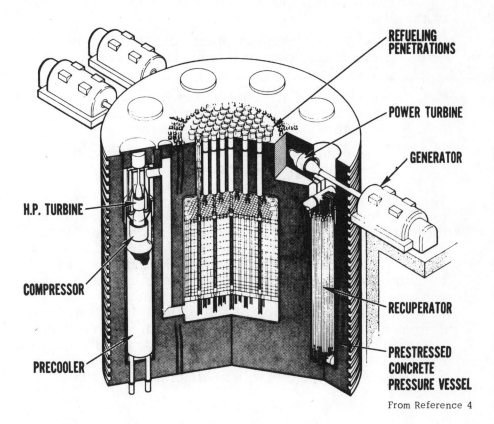

REFUELING PENETRATIONS

POWER TURBINE

GENERATOR

H.P. TURBINE

COMPRESSOR

PRECOOLER

RECUPERATOR

PRESTRESSED CONCRETE PRESSURE VESSEL

From Reference 4

FIGURE 12.3
Large High-Temperature Gas-Cooled Reactor with Helium
Closed-Cycle Gas Turbines for Power Generation. Heat
is Rejected at the Precoolers   through intermediate
water loops to the ambient Air.

The following advantages are claimed by the proponents of the plant:

> 10-15% reduction in plant cost, excluding cooling system, over the same size Rankine cycle HTGR plant.
>
> Reduction of air-cooling equipment cost by about two-thirds for the helium turbine plant, were they both to be air cooled.
>
> Lower susceptibility to higher ambients than air-cooled LWR plants.

The elimination of the large steam generators, steam turbine, feedwater heaters, pumps, etc., all required for a steam cycle plant, will result in a compact plant arrangement as shown in the artist's concept, Figure 12.3. Three independent power producing loops with individual electrical generators are shown, the helium compressors powered by separate turbines.

A pictorial comparison of the natural draft tower requirements of a helium turbine, air-cooled HTGR plant; a Rankine cycle, air-cooled HTGR plant; and Rankine cycle HTGR plant with a wet cooling tower, all of approximately the same size and efficiency, is shown in Figure 12.4.

12.4 Other Gas-Cooled Reactors with Gas Turbine Cycles.

The arguments for the helium turbine, air-cooled HTGR plant does not apply to the proposed demonstration Gas-Cooled Fast Breeder Plant* where helium will again be the reactor coolant, as in this case the ceramic fuel elements will be enclosed in cladding, limiting maximum helium temperature to below 1300°F, which will be too low for an efficient gas turbine cycle, Future GCFB plants without such limitations, will probably be able to utilize the direct gas turbine cycle rejecting heat to air.

*Electrical World, June 1, 1971; p27.

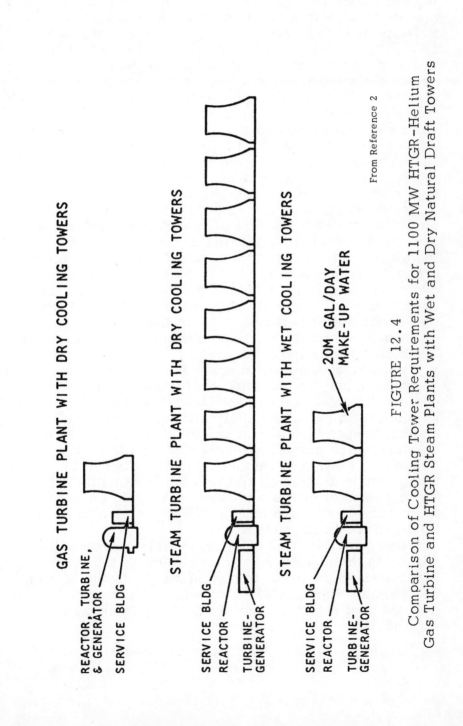

FIGURE 12.4

Comparison of Cooling Tower Requirements for 1100 MW HTGR-Helium
Gas Turbine and HTGR Steam Plants with Wet and Dry Natural Draft Towers

It is of some interest that the first air-cooled nuclear
reactor plant with a gas turbine cycle was built and
initially operated over a decade ago - in September
1962. The ML-1, a 300 kWe mobile power plant for
the U. S. Army, utilized a 3300 kw thermal, hetero-
geneous-water moderated, gas-cooled reactor. The
reactor coolant and working fluid for power generation was nitrogen (with 0.5% oxygen), entering the
gas turbine at about 300 psig and 1200°F and expanded to a ratio of 2.5. Heat rejection to air, at about
500°F, was directly from the cooling-working fluid
loop across an all-aluminum, fan-cooled, crossflow
heat exchanger with perforated fins on both sides.

A closed gas turbine cycle with heat rejection to
air is, of course, possible also with the Advanced
Gas-Cooled Reactor (AGR) that utilizes carbon dioxide as coolant. But in this case, the lower gas temperature leaving the reactor will considerably reduce
cycle efficiency. In a closed gas turbine cycle, carbon dioxide will have about half the overall heat transfer coefficient of helium and about the same as air,
which will result in increased surface for the numerous
cycle heat exchangers.

The more efficient with lower inlet temperatures
carbon dioxide condensation cycle, discussed in the
chapter "Special Fluid Power Plants", will not be
suitable for air-cooling applications, as the high
critical temperature of carbon dioxide, 87.8°F, will
eliminate such plants from consideration in most
climates. Certain advantages can be obtained by
using carbon dioxide as the working fluid with Fast
Breeder Reactor plants with low peak cycle temperatures when low heat rejection temperatures permit
liquid phase compression.

Other possible high efficiency nuclear cycles discussed in the chapter mentioned, such as the binary
potassium-steam cycle in conjunction with the Molten
Salt Reactor Plant (MSR), also offer the prospect of
reduced heat rejection and should be amenable to air
cooling.

REFERENCES

1. Goodjohn, A.J. and Fortescue, P.; Environmental Aspects of High-Temperature, Gas-Cooled Reactors; Proceedings of the American Power Conference, Vol. 33, 1971.

2. Fortescue, P.; Gas Turbines and Nuclear Power; Combustion, December 1972.

3. Bell, F. R. and Koutz, S. L., Gas Turbine HTGR and the Environment; ASME Paper 72-WA/NE-8. Also in Combustion, April 1973.

4. Krase, J.M., Morse, D.C. and Schoene, D.E., The Direct-Cycle Nuclear Gas Turbine with Economical Dry Air Cooling; Proceedings of the American Power Conference, Volume 34, 1972.

5. Bammert, K. and Bohm, E.; Nuclear Power Plants with High-Temperature Reactor and Helium Turbine, ASME Paper 69-GT-43.

6. Bammert, K.; The Nuclear Power Gas Turbine; Paper 2.4-96, The Eighth World Power Conference, Bucharest, 1971.

7. Keller, C.; The Nuclear Gas Turbine; Gas Turbine, July-August 1965.

8. Keller, C.; The Escher Wyss-AK Closed Cycle Turbine, Its Actual Development and Future Prospects; ASME Transactions, 1946.

9. Nakatani, R.E., Miller, D. and Tokar, J.V.; Thermal Effects and Nuclear Power Stations in the USA; Paper Presented at the International Atomic Energy Symposium on Environmental Aspects of Nuclear Power Stations, August 1970, UN Headquarters, New York, New York.

0. Christ, A.; Heat Transfer and Pressure Drop; Escher Wyss News, January 1966.

1. Gasparovik, N.; Nuclear Helium Gas Turbines; Energy International, April 1971.

REFERENCES (Cont'd.)

12. Janis, J.M. and Seeley, G.T.; The Development and Operating Experience of the ML-1 Mobile Nuclear Power Plant, ASME Paper 66-GT/CLC-12.

Chapter Thirteen

SPECIAL FLUID POWER PLANTS

13.1 The Search for a Superior Working Fluid - and Its
      Attributes

The suitability of various fluids to replace steam or
share the work of expansion in binary fluid cycles has
been of interest from the early times of piston steam
engines to the present day of nuclear reactors, space
mission power plants, and the search for a non-pollut-
ing automotive engine.

It is reported that early in the 19th century Davy was
the first to suggest the use of a fluid that would vapor-
ize and expand, performing further work by the heat rejected from a steam turbine. At the time steam engines
exhausted to near atmospheric pressure and corresponding high condensing temperatures, considerably above
the ambient. By 1848 several marine engines using
water as the primary working fluid and ether as the
secondary fluid vaporizing from the heat rejected by
the steam were constructed by the French engineers
Du Trembley and Bourdon, the latter also the inventor
of the synonymous pressure gauge.

More recent, and of practical interest, were the efforts of W. L. R. Emmet and the General Electric Company in developing a binary mercury-steam cycle power
plant of which about half a dozen were built between
1928 and 1950 in ratings up to 20 MW, with about 40%
of the output derived from the mercury turbine. The development of a mercury boiler was the main obstacle
overcome in constructing the plants, as mercury does
not wet steel, with resultant poor heat transfer and
tube burnouts. The higher efficiency, realized because of the higher average temperature at which heat
was added to the cycle, was the incentive in developing the plants, notwithstanding the high toxicity of
the mercury vapor, but parallel improvements in the
steam cycle removed this incentive.

Very briefly stated, lacking a single fluid with
superior properties to replace steam, the search has
been for a cheap, stable, non-toxic, non-corrosive
fluid with higher critical temperature and lower critical pressure for topping, or superposed, cycle appli-

cations, so that heat can be added at high temperature
without the need for high strength materials at the elevated temperature. An appropriate lower critical temperature and pressure fluid, with lower specific volume
than steam and near atmospheric pressure at available
ambient condensing temperatures is also needed for
bottoming, or subposed, cycle applications. Among
other desirable properties for both fluids are large
latent heat and low liquid specific heat - or a near
vertical saturated liquid line - so that most of the
heat is added during change of phase without the need
for the complexity of regenerative feedheating for "car-
notizing the cycle"; and a near vertical saturated vapor
line so that little moisture results during expansion,
and conversely avoiding the need for condensing a
superheated vapor if the expansion ends in the superheated region.

The undesirable properties of steam are the extreme
pressure variation between heat addition and heat rejection (from 3500-2400 psia to as low as 0.5 psia),
which create problems with piping and air inleakage,
and the low specific volume at heat rejection temperatures which result in large size exhaust ends and considerable leaving losses; while desirable steam properties are the large latent heat and the relative incom-
pressibility of the water that minimizes boiler feed
pump work. The saturation pressure-temperature relation and the critical point for several fluids are shown
in Figure 13.1.

13.2 Two Criteria: Improved Efficiency or Reduced
Investment in Equipment

The first three references to this chapter are recommended for a further introduction, as only the air-
cooled aspects of special fluid power plants will be
elaborated upon here. They can be identified as those
aspects that reduce heat rejection by improving plant
efficiency and those that simplify turbine exhaust end
and air-cooled heat exchanger design because of low
specific volume and near ambient pressure at condensing temperatures, thus better utilizing variable and low
ambients or eliminating freezing problems on account
of a low freezing point.

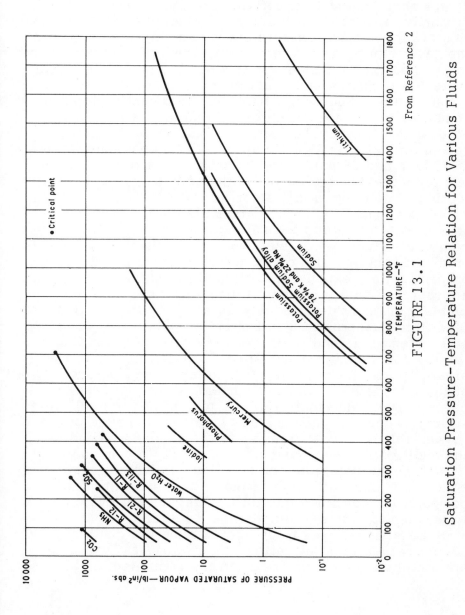

FIGURE 13.1

Saturation Pressure–Temperature Relation for Various Fluids

Improving efficiency in fossil-fired plants by super-position has been unsuccessful but for the short time span General Electric experience mentioned above. Primarily, fire side corrosion has hindered the introduction of higher cycle temperatures*, as certainly has been the case with steam. The current stress on clean fuels for air pollution considerations will also reduce fire side corrosion problems and could encourage higher cycle temperatures. Such an example is the recently proposed combined cycle of a gas turbine in parallel with a binary potassium-steam cycle, with combustion gases to the gas turbine at 1700°F and vapor to the potassium turbine at 1540°F and 29 psia, from a common gas turbine combustor/potassium boiler and claiming half the heat rejection of a conventional steam plant.

As to supposition cycles, considerable literature exists as to the purported savings in conventionally cooled plants to be realized when the multiple low-pressure sections of a steam turbine--large and disproportionately expensive compared to output items-- are replaced with a compact refrigerant turbine. Unfortunately a recent detailed design study by the Central Electricity Generating Board of Great Britain that considered ammonia and freon 21, the two most promising fluids, concluded that investment will be higher for the binary plants, as compared to a steam plant, on account of the intermediate heat exchanger and the increased complexity of the binary fluid plant. Efficiency will also suffer, except for partial load operation, because of the irreversibility of heat transfer to the secondary fluid and the additional power input into the cycle in pumping the relatively compressible fluids. The CEGB study also concluded that, although faring better than in the above comparison, combined gas turbine-refrigerant cycles were inferior to gas turbine-steam cycles.

Supposition cycles to utilize extreme low ambients have received little attention so far as most power generation has been concentrated in the populated, mild

---

*There are indications that fire side corrosion with coal-fired boilers is limited to metal (tube) temperatures between 1100°F to 1300°F.

climates with cooling from moderate temperature waters.
It is of interest that a recently announced 750 kw freon
power turbine has been developed by a Siberian technical institute. Another 3800 kw fluorocarbon turbine has
been announced by a Japanese firm, to utilize waste
heat for power generation, or alternately to drive a refrigeration compressor.

13.3 The Ammonia Bottoming Cycle

A recent study by a non-profit research laboratory for the
Office of Coal Research, Department of the Interior, that
considers large mine mouth plants in arid regions to
utilize western coal, strongly favors a binary steam-
ammonia cycle, the ammonia condensing directly inside
air-cooled heat exchanger tubes. Utilizing wide ambients
is mentioned as a useful feature of the plant, but primarily a cost advantage of over a third of the surface and the
cost of the air-cooled heat exchanger is claimed, osten-
sibly on the basis of a much higher condensing coefficient
for ammonia over the waterside film coefficient of the
Heller system*. The dominant air side coefficient inferred
from other figures in the report was about the same in both
cases.

*A 40% improvement in overall heat transfer coefficient
by condensing ammonia is claimed, from 5.7 Btu/hr.-ft.^{2}-
°F for an indirect system heat exchanger (based on outside
bare tube surface) to 8.0 Btu/hr.-ft.^{2}-°F when condensing ammonia. The report compares the condensing coefficient of ammonia inside tubes - stated as between 1200-1500 Btu/hr.-ft.^{2}-°F to an inside tube film
coefficient for water of 400, for an air-cooled indirect
system heat exchanger of 1 inch OD, 18 BWG tubes. Assuming average water temperature at 120°F, the above
coefficient will correspond to a water velocity of about
1.2 fps. Both Rugeley and Ibbenburen have tubes of
about .6 inches ID and water velocity of about 4 fps,
which will result in an inside film coefficient of about
1,100 Btu/hr.-ft.^{2}-°F.

The ammonia-to-air heat exchanger will actually require more surface than that of the indirect system and
will have to be considerably stronger to withstand the
high condensing pressure of ammonia (300 psia at 123.2
°F). Other disadvantages of the steam-ammonia plant
will be the high wetness and erosion problems of the
ammonia turbine, ammonia toxicity and corrosiveness,
and the high feedpump work input to the cycle because
of the greater compressibility of ammonia compared to
water and high ammonia flow. (Two pounds of ammonia
have to be expanded, and subsequently pumped, to ob-
tain the same output as from expanding one pound of
steam.)

The impression should not be formed that a steam-
ammonia air-cooled cycle is without merit. Heller has
suggested the use of the steam-ammonia cycle, over
other refrigerants that were considered, for plants of
large output and in very cold climates, where in the
latter case steam at any rate has to be displaced from
the lower part of the expansion line to avoid freezing
problems. Problems of variable ambients and low load
operation, usual with air-cooled steam plants, will
also be simplified, as can be inferred from Figure 13.2,
where the specific volumes of saturated steam and saturated ammonia for various condensing temperatures
are plotted.* The extreme variation in steam specific
volume results in turbine exhaust end design difficulties, as the same turbine exhaust end area cannot efficiently accommodate widely varying volume flows.
The paper by Heller is valuable for the many insights
it offers into the problems of future power, large plant
generation and the controlling factors, as well as for
the analysis of binary cycle design complexities and

*The lower specific volume of ammonia compared to
steam should not be misconstrued as implying a smaller
air-cooled condenser, as the heat transfer surface requirements and not vapor volume dictate heat exchanger
size. Heat transfer surface is determined mainly by the
air side parameters--bare tube to fin ratio and air veloc-
ity--the cooled fluid inside the tube having little effect
on the overall heat transfer coefficient.

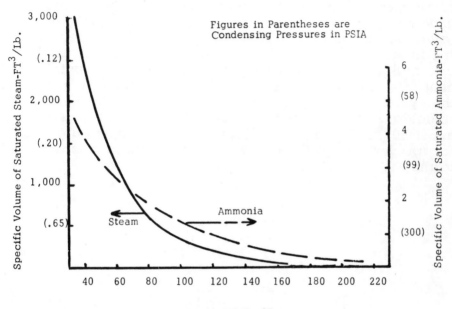

FIGURE 13.2

Specific Volumes of Saturated Steam and Ammonia

operational problems. The second reference given at
the end of the chapter, with the many contributed dis-
cussions, indicates the variance of opinion and strongly held views on the subject of binary cycle plants,
while the fourth reference demonstrates how in a detailed design study, in this case for the CEGB, many
of the nebulous theoretical benefits can evaporate
when the necessities of hardware design have to be
taken into account.

13.4 The Carbon Dioxide Liquid Compression Cycle; Potassium and Helium Cycles.

Another cycle of interest that has received considerable attention abroad, particularly in the USSR, is the liquid compression carbon dioxide cycle that combines features of the gas turbine cycle with the low work input of the Rankine cycle that results from condensation during heat rejection and pumping of the collapsed - in-volume working fluid. Since carbon dioxide has a critical temperature of 87.8°F (and a critical pressure of 1071.1 psia); heat rejection at low temperature is a requirement and the cycle could have applications for air-cooled plants in cold climates where the ambient air does not exceed about 60°F. Efficiency vs. initial cycle temperature for two variants of the carbon dioxide cycle is compared to the steam in Figure 13.3.

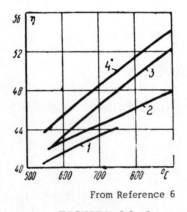

From Reference 6

FIGURE 13.3

Variation in net efficiency of steam/water cycle and high capacity CO_2 plants with increase in the initial temperature. 1 - steam/water cycle plant with one reheat; 2- steam/water cycle plant with two reheats[9]; 3 - carbon dioxide plant with two-phase compression; 4- carbon dioxide plant with split flows.

Desirable features of the carbon dioxide liquid compression cycle are the following:

> Higher average heat addition temperature to the
> cycle, as there is no upper temperature limit to
> regenerative heating, the condensate liquid com-
> pressed to above critical pressure vaporizing
> without change in phase, mostly in a gas-to-gas
> heat exchanger similar to a gas turbine regenera-
> tor, from heat transferred from the expanded gas.
> In the case of the steam cycle, the saturated
> temperature, at the given pressure, determines
> the upper regenerative feedwater heating temp-
> erature, and the amount of heat added from an
> external source at top cycle temperature is re-
> duced, even with reheat, which becomes un-
> necessary in the carbon dioxide cycle as no
> moisture forms during expansion.

> Fewer turbine stages and more compact turbines.

> The carbon dioxide cycle utilizes lower ambients
> more efficiently, some of the heat previously re-
> jected being utilized for regenerative feedheating,
> as shown in Figure 13.4.

The carbon dioxide cycle has also received attention
for binary fluid plants in the USSR and for sodium-cool-
ed and gas-cooled fast breeders in Europe. In the latter
case, considerable simplicity results by replacing the
intermediate sodium loop and the sodium-to-water and
steam heat exchangers and steam system with a carbon
dioxide power cycle which receives heat directly from
the reactor's primary sodium coolant. The desirable
nuclear and physical properties of carbon dioxide, as
compared to the complexities of a sodium-water system,
make such an arrangement attractive in spite of its lower efficiency at the temperatures currently considered,
but any considerations of air cooling for such plants is
contingent upon the year-round availability of cold
ambients.

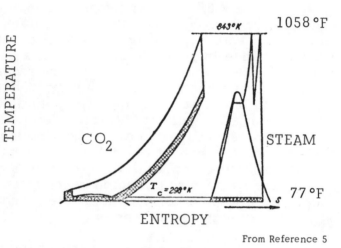

From Reference 5

FIGURE 13.4

Increase in useful work (shaded areas) performed in
the carbon dioxide and steam cycles on reducing condensing temperature. The entropy scale for the carbon dioxide cycle has been increased by about 3.5
times to show equal work areas for the two cycles.

It should be kept in mind that with the current low
nuclear fuel costs and the prospects for even lower
cost nuclear fuel from breeders, cycle efficiency considerations are not overriding in nuclear plants, as the
marketing success of the Light Water Reactor plants
makes abundantly clear. Nonetheless, binary nuclear
cycles for improved efficiency have been proposed -
from a suggestion for a sulfur-steam cycle receiving
heat from a boiling sulfur reactor to an earlier version
of the potassium-steam cycle mentioned previously
for a combined-cycle arrangement. In the earlier
nuclear version the potassium received the heat from a
Molten Salt Reactor, the reactor concept that has re-
peatedly attracted mention for its projected low fuel
cost, breeding potential and simple, process type fuel

handling. Potassium vapor wetness and shaft sealing
may present problems, but a small potassium turbine
has been tested in conjunction with space power plant
applications. A large turbine manufacturer, together
with a boiler manufacturer and a consulting engineering company, are seeking Federal support to evaluate
the binary potassium-steam cycle for large fossil
power plants; the cycle will also benefit from the parallel development of large liquid sodium components
for the LMFB reactor.

The need to limit heat rejection may provide the impetus for improving nuclear plant efficiency, as in the
closed helium gas turbine cycle, amenable to air cooling, and proposed in conjunction with the High-Temperature Gas-Cooled Reactor that is discussed in some
detail in Chapter 12.

REFERENCES

1. Meyer, C.A. and Fischer, F.K., Working Fluids for
    Power Generation of the Future; Proceedings of the
    American Power Conference, Volume XXIV, 1962. Ex-
    cerpted in the "Westinghouse Engineer," January 1963.

2. Wood, B., Alternative Fluids for Power Generation;
    Proceedings of the Institution of Mechanical Engineers,
    Volume 184, Part 1, No. 40, 1969-1970.

3. Bidard, R., Novel Thermodynamic Cycles and Fluids
    (in French); Revue Generale de Thermique, Vol. IX,
    No. 99, March 1970.

4. Eaves, P.S.K. and Hadrill, H.F.J., Factors Affecting
    the Application of Binary Cycle Plant to the C.E.G.B.
    System; Paper C1-241, VII World Power Conference,
    Moscow 1968.

5. Gokhshtein, D.P., Smirnov, G.F. and Kirov, V.S.,
    Some Features of Binary-Cycle Systems with Non-
    Aqueous Vapours; Thermal Engineering (Teploenergetika),
    January 1966, HVRA Translation.

6. Dekhtyarev, V.L., Problems of Thermodynamic Analysis of
    Actual Cycles in Power Generating Plants; Thermal Engin-
    eering (Teploenergetika), December 1967, HVRA Transla-
    tion.

REFERENCES (Cont'd.)

7. Heller, L., New Power Station System for Unit Capacities in the 1000 MW Order; Acta Technica Hungarica, 1965.

8. Slusarek, Z.M., "The Economic Feasibility of the Steam-Ammonia Power Cycle"; Franklin Institute Research Laboratories Report prepared for the Department of the Interior, PB184331, 1968.

9. Angelino, G., Carbon Dioxide Condensation Cycles for Power Production; Transactions of the ASME, Journal of Engineering for Power, July 1968.

10. Casci, C. and Angelino, G., The Dependence of Power Cycles' Performance on Their Location Relative to the Andrews Curve; ASME Paper 69-GT-65.

11. Sawle, D.R. and Salisbury, J.K., A Binary-Vapor Nuclear Power Plant, ASME Paper 60-WA-309.

12. Fraas, A.P., A Potassium Steam Binary-Vapor Cycle for a Molten-Salt Reactor Power Plant; Transactions of the ASME, Journal of Engineering for Power, October 1966.

13. Fraas, A.P., A Potassium Steam Binary-Vapor Cycle for Better Fuel Economy and Reduced Thermal Pollution; ASME Paper WA/ENER-9; also, How About a Potassium Topping Cycle? ; Electric Light & Power, March 1972.

14. Wilson, A.J., Space Power Spinoff Can Add 10+ Points of Efficiency to Fossil-Fueled Power Plants. 7th Intersociety Energy Conversion Engineering Conference, San Diego, 1972.

15. Katterhenry, A.A., Gas-Turbine Nuclear Power Plants; Reactor Technology, Volume 13, No. 1, Winter 1969-1970.

16. Fluorocarbon Turbine; News item in Mechanical Engineering, January 1971.

17. Kahaeb, A.A. et al, Large Capacity Water-Freon Power Installations; Paper C1-10, VII World Power Conference, Moscow, 1968.

REFERENCES (Cont'd.)

8. Properties of Commonly Used Refrigerants; Air Conditioning and Refrigerating Machinery Association, Inc., Washington, D.C.

9. Steigelmann, W.H.; Alternative Technologies for Discharging Waste Heat, in "Power Generation and Environmental Change"; Berkowitz, D.A. and Squires, A.M. editors, M.I.T. Press, 1971.

0. Steigelmann, W.H., Seth, R.G., and Watchell, G.P.; Binary-Cycle Plants Using Air-Cooled Condensing Systems; Proceedings of the American Power Conference, Volume 34, 1972.

1. Chermanne, J.; Nuclear Power Plant Incorporating Low-Pressure CO_2-Gas Turbine Cycle (in German); Brennst.-Warne-Kraft, September 1971.

Chapter Fourteen

SYSTEM PLANNING CONSIDERATIONS

14.1 Past and Present of System Planning Studies

System planning, which refers to planning generating
plant additions and the associated transmission lines
to meet expected future demands, usually within the
constraints of minimum overall cost for a desired reliability level, was until fairly recently a straight
forward engineering/economic analysis, based prin-
cipally on future load projections, equipment and
fuel economics and system reserve requirements. For
a projected increase in demand, the economics of size
and efficiency improvements with advancing steam conditions dictated the size and cycle arrangement for the
new generating plant that was to operate base load,
while older units were relegated to lower load factor
operation and eventually peaking and reserve status
prior to retirement. The other major decision-that of
site selection-was made on the basis of proximity to
the load, fuel transportation facilities and a source of
cooling water for the plant's condenser.
    In the last few years great uncertainties have began
to beset the formerly simple engineering/economic analysis. Such parameters as system reserve requirements,
fuel costs, etc., that had well defined and fairly constant values for each utility system, based primarily on
size and location of the utility, now become confused
and sometimes unpredictable. Highlighted by but pre-
ceding the Northeast Blackout of 1965, the question of
adequate reserves for large interconnected systems was
raised. Following the blackout and spurred by continuing
delays in the construction of large units, a deluge of gas
turbine orders has resulted in gas turbine plant of 15-25%
of the total of some systems from a previously miniscule
or non-existent portion. Also delays during construction,
coupled with rising equipment and labor costs, have made
interest during construction (IDC) a substantial part of a
power plant's cost.
    The lack of stability in fossil fuel prices-coal prices
nearly doubling in a period of five years and highly er-
ratic oil prices-coupled with demands for not readily

available low sulfur fuel supplies, have clouded the
fuels market and made the fixed-price, long-term fuel
contract a thing of the past. Variable tax and depreciation treatment of plant investment, with changing guidelines, has also introduced the accounting aspect into
planning considerations.

More recently air pollution and thermal pollution
considerations arising from ecological and societal
reservations that challenge the well promoted and ever
increasing appetite for electricity consumption are resulting in regulatory guidelines that are currently in
the process of formulation and transition and often over-
shadow strictly engineering/economic considerations.

The plant available to utility planners, to meet the
rising demand for electricity while conforming to the increasingly diverse requirements, has also undergone
startling change. Concepts unfamiliar to power plant
engineers in the nineteen fifties now are the subject of
newspaper reporting. The gas turbine and pumped storage plants are helping many utilities meet the daily system peaks. Large cooling towers and long transmission
lines have brought mine-mouth plants and their mixed
benefits to mining communities, far from the population
centers they serve. And, of course, there has been a
proliferation of public hearings and nuclear plants
stalled in various stages of completion. Specific type
of plant is considered for meeting fairly well-defined
peaking, intermediate and baseload demand. System
planning techniques have also changed: from simple
calculations to compare the projected cost of a few
alternate expansion schemes to computer-oriented
probabilistic and dynamic programming methods and to
planning with the assistance of computer corporate models.

14.2 Generating Plant Options and Their Overall Effect
      on the Cost of Electricity to the Consumer

As already mentioned, a utility system's generating
plant is selected on the basis of meeting baseload,
intermediate and peaking demand and providing adequate reserves at a minimum overall cost. Baseload
plant is usually high in first cost but of high efficien-

cy and low fuel and maintenance cost. Baseload operation justifies the high investment as it can be written off over the many kilowatt hours generated per year;
the same criterion requires the high efficiency and low
fuel and maintenance costs. Nuclear plants of course
fall in this category, together with large, efficient fossil-fired plants and hydro plants, although few sites for
the latter remain undeveloped. For the other operating
extreme-peaking and reserve requirements-low first
cost plant is preferred, as the investment has to be ap-
portioned over the few kilowatt hours generated by this
type of plant while efficiency, fuel and maintenance
costs are of secondary importance. Quick starting a-
bility is also a requirement as the plant is subject to
frequent and often unscheduled startups. Gas turbines
fall in this category, as does the pumped storage plant,
where the system's topography makes the latter feasible.

The requirements for plant to satisfy intermediate
duration load fall between the two extremes-a compromise between investment and the so-called operating
costs. Older plants usually fulfill this need in many
systems, while the simpler cycle, lower investment
and efficiency steam plant is currently gaining acceptance for this service. The C.E.G.B. in Great Britain
and the USSR have apparently opted for combined cycle
plants to meet mid-range demand and are planning for
their widespread introduction, while their acceptance
in this country, after meeting resistance for several
years, is now gaining momentum.

It is difficult to generalize in dividing the load duration curve into peaking, intermediate and baseload
demand, as these terms are meaningful only in a functional context and are specific only to a utility system
at a given time. They depend on the load factor, nature of the load, relation of daily, weekly and seasonal peaks, etc., but perhaps peaking demand can be assumed to extend to up to 1,000 hours per year, mid-
range to 4,000 hours per year, with the remainder
divided between intermediate (to 6000 hours) and normal baseload.

Figure 14.1 shows a simplified form for presenting
and comparing total annual costs of owning and operating different types of generating plant at varying

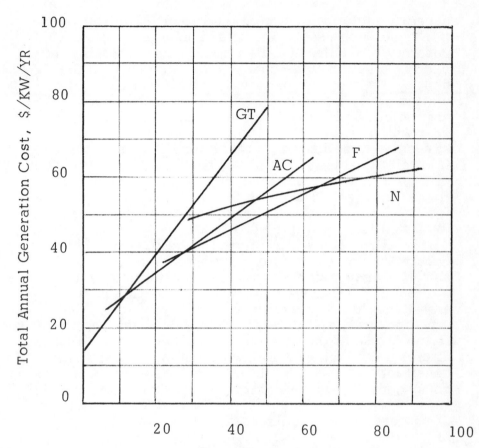

FIGURE 14.1
Owning and Operating Costs for Various Types of Plant

G.T.   Gas Turbine at $90/KW, 14,000 BTU/KWHR H.R.,
90 c/MMBTU fuel, 2.5 mills/KWHR O&M cost.

A.C.   Air-Cooled Steam Peaking/Cycling plant at $140/KW
1.0 mil /KWHR O&M cost.

F   Coal-Fired Steam plant at $180/KW, 9,000 BTU
KWHR H.R., 50 c/MMBTU fuel, 1.0 mill/KWHR
O&M cost.

N   Nuclear L.W.R. plant at 260/KW, 10,000 BTU/
KWHR H.R., 19 c/MMBTU fuel at 80% L.F.
and 25 c at 25% L.F., 1.2 mills/KWHR O&M cost.

15% F.C.R. in all cases.

capacity factors; plant cost data for 1972 initial operation have been used. The total annual cost includes a
capacity cost, the installed cost of the generating capacity times the fixed charge rate, which is independent of the hours of operation, a maintenance and operating cost that varies somewhat with the hours of operation but is usually assumed a constant, and a fuel
cost which is directly proportional to the hours of operation for fossil fired plants; hence the near straight-
line relationships. The slight curvature for the nuclear
plant is due to the variable nuclear fuel cost that reflects core capital charges for varying residence of the
fuel in the reactor. Another form of presenting such information is shown in Figure 11.2.

The above costs are referred to as busbar or total
generation costs. Before the power reaches the consumer, transmission lines and a distribution system
are required. Transmission and distribution costs are
also items highly specific to a utility system, whether
urban or rural, and its state of development. Table 14-Ia
is indicative of the relative importance of such expenditures on an average nationwide basis of total utility
construction expenditures. Table 14-Ib shows the average
component costs of electricity to the consumer in 1969;
it includes fixed charges on the investment and operating costs, i.e., fuel, maintenance, etc.

	TABLE 14-Ia* Investment In Facilities 1965-69	TABLE 14-Ib* 1969 Component Cost of Electricity
Generation	41.9%	50%
Transmission	21.2	13
Distribution	32.5	37
Other	4.4	

*Table 14-Ia is based on Table 50S and Chart VIII-B in
"Construction Expenditures-Investor Owned Electric Utilities, 1965-1969 Inclusive", EEI Statistical Yearbook for
1969. Table 14-Ib is derived from Table 19.1, Part I, of
the 1970 National Power Survey.

Unlike the choice of diverse generating plant, little trade-off can be made between investment and operating costs for transmission and distribution facilities. As a result the penalty for intermittent transmission and distribution of power is more severe than for intermittent generation, and the higher cost of intermittent consumption is of course reflected in utility rates.

The overall effect to the consumer's cost of power can be seen from a simplified example. Assuming that a system is generating 10% of its yearly output from air-cooled condenser plants at 10% higher busbar cost than the average for the remainder of the system, the increased cost to the consumer for a system with a 50% generating cost component will be .5%. The example assumes that no reduction in other costs, such as transmission, results from the selection of air-cooled condensing plants

14.3 The Case for Baseload Air-Cooled Power Plants

Air cooled condenser plants have been considered in the past only for baseload duty. Table 14-II summarizes the conclusions of the major published studies on the effect of air cooling on total generation costs.

The conclusions of the study shown last in the table appear to be surprisingly within the range, considering that the study preceded the others by nearly a decade.

The preference for baseload air-cooled condenser plants is based on the arguments that the extra costs for the air-cooled condensing system and the loss in efficiency resulting from the higher back pressure can be made up through savings in transmission line construction and through eliminating the delays and reducing interest during construction, assuming that the alternate plant was to be once through cooled at a distant site and require a variety of permits and modifications for discharging thermal effluents. When cooling water is scarce, air-cooled plants near fuel sources will be justified on the basis of savings in fuel transportation costs. Evaporative cooling towers, introduced a decade ago with similar arguments, require large consumptive usage of water, which is not always

TABLE 14-II
POWER GENERATION COSTS WITH
AIR COOLED CONDENSER PLANTS
AS COMPARED TO PLANTS WITH WET COOLING TOWERS
REFERENCE A[1]

LOCATION	VARIOUS	VARIOUS
SIZE, MW	750 MW — @ 3.5' Hg	500 MW
TYPE	FOSSIL - INDIRECT	NUCLEAR - INDIRECT
YEAR	1970	1970
DESIGN DRY BULB(MAX.), °F	100°F	100°F
DESIGN ITD, °F		
TURBINE BACK PRESSURE,		
INCHES HG.	8.5 -15.6	8 -15.6
PLANT LOAD FACTOR, %	60	80
FIXED CHARGE RATE, %	15	15
FUEL COST, ¢/MMBTU	15,30,40	15
SUMMER REPLAC. POWER,		
$/KW	0-100	0-100
ADDITIONAL EQUIPMENT	$15.40-$38[2]	$26.50-$51[2]
COST, $/KW		
TOTAL ADDITIONAL		
BUSBAR COST, MILLS/KWHR		
BREAKEVEN TRANSMISSION		
LINE LENGTH, MILES	75-200	130-250
BREAKEVEN EVAP. WATER		
COST, $/1000 GALLONS	.75-1.90	.70-1.40
BREAKEVEN FUEL OR		
TRANSP. COST, ¢/MMBTU	5-12	

[1] SEE STUDIES LISTED AT THE END OF THE REFERENCES FOR THIS CHAPTER.

[2] CAPITALIZED OWNING AND OPERATING COST DIFFERENTIAL

TABLE 14-II (cont.)

REFERENCE B		REFERENCE C
VARIOUS	VARIOUS	EASTERN, S.W. & WESTERN
800 MW-@ 3.5" HG	800 MW	860 to 928(@ 3.5" HG)
FOSSIL—INDIRECT	NUCLEAR—INDIRECT	NUCLEAR—INDIRECT
1970	1970	1975
FIG. SHOWN ARE FOR CHICAGO PLANT		FIG. SHOWN ARE FOR THE
96 (#0 HOURS)	96	860 NW EASTERN PLANT
55-61	57-67	56°-67°
75	75	
8-18	8-18	12,15,18
25-40	11-20	15,18,21
100	100	100
17-20 plus	23-27 plus	16-25 plus
6-8 for G.T.	13-14 for G.T.	9-12 for G.T.
.48 (7-10%) with		.72-1.07 (10-14%)
61°ITD,15% FCR,35¢ FULL		
		70-100
.75		.85-1.25
5		

TABLE 14-II (cont.)
POWER GENERATION COSTS WITH
AIR COOLED CONDENSER PLANTS
AS COMPARED TO PLANTS WITH WET COOLING TOWERS

LOCATION	REFERENCE D PHILA. @ PHOENIX	REFERENCE E NORTH DAKOTA
SIZE, MW	959 @ 104°F	185 (@ 3.5" HG)
TYPE	NUCLEAR-SURFACE CONDENSER	COAL-DIRECT
YEAR	1971 & 1978 FIGURES SHOWN ARE FOR PHILA. IN 1971	
DESIGN DRY BULB(MAX.), °F		75°F @ 5000' ELEV.
DESIGN ITD, °F		
TURBINE BACK PRESSURE, INCHES HG.		5"
PLANT LOAD FACTOR, %	70	70
FIXED CHARGE RATE, %	15	15
FUEL COST, ¢/MMBTU		17[4]
SUMMER REPLAC. POWER, $/KW	100 & 10 MILLS/KWHR FROM 959 to 1132 MW	NO
ADDITIONAL EQUIPMENT COST, $/KW	24.70[3] plus 12.20 for G.T.	13
TOTAL ADDITIONAL BUSBAR COST, MILLS/KWHR	17%	.4(55%)
BREAKEVEN TRANSMISSION LINE LENGTH, MILES	317	
BREAKEVEN EVAP. WATER COST, $/1000 GALLONS	1.07	
BREAKEVEN FUEL OR TRANSP. COST, ¢/MMBTU		

[3] WITHOUT DEDUCTING THE COST OF AN EVAPORATIVE COOLING SYSTEM

[4] AN 8¢/MMBTU FUEL TRANSPORTATION DIFFERENTIAL HAS ALSO BEEN CREDITED
TO THE AIR-COOLED PLANT

TABLE 14-II (cont.)

REFERENCE F GERMANY	REFERENCE G DRY BELT OF U.S. 200 MW
FOSSIL	COAL - INDIRECT 1963
	95° (5%)
	45 & 55
	6.7 & 7.6
57	75(1st DECADE)
16	13.5
72	20 ESCAL. to 35
NO	100
7.60	11.30-18
.42	.4-.6(8-12%)
	50-70
.90	.49-.73
5-6	$3-$4/TON @ 100 MILES

NOTE: IN ORDER TO PRESENT INFORMATION IN A COMPARATIVE FORMAT, SOME OF THE FIGURES SHOWN IN THE TABLE ABOVE WERE DERIVED ON THE BASIS OF RELEVANT MATERIAL IN THE REFERENCES INDICATED AND MAY NOT REFLECT THE INTENTION OF THE AUTHORS. RESORT TO THE ORIGINAL SOURCE IS RECOMMENDED IF ANY OF THE INFORMATION IN TABLE 14-II IS TO BE FURTHER UTILIZED.

available and can have potentially undesirable environmental effects by promoting fog and cloud cover and often require regulatory approval similar to the once through
plant for consumptive water usage from streams, and for
disposal of the high in solids and chemicals blowdown.
Generally, then, the argument is that savings in transmission line, fuel transportation costs, water usage,
lead time and I.D.C., individually or in combination,
will benefit the air-cooled condenser plant; while environmental benefits will result by eliminating thermal
effluents, evaporative plumes, or salt drift from towers
using saltwater as makeup. The specific costs of the
tangible items vary widely from utility to utility, but
often breakeven figures such as miles of transmission
line, cost per thousand gallons of consumptive water
usage or fuel savings in cents per million BTU, are
given that will offset the total additional generation
cost of the air-cooled plant.

Cost of transmission facilities vary greatly depending on the amount of power transmitted, voltage and
length of the transmission line, and the cost of land
and rights of way. Figure 14.2 indicates the general
cost trends for the assumptions shown. Figure 14.3
compares the average costs of energy transportation
prevailing in the Western United States in the sixties
by transmission line or as a fuel. In the
latter case a heat rate of 10,000 BTU/KWhr has been
assumed in converting the fuel into electrical energy.
Figure 14.4 indicates the geographical distribution of
potential adverse environmental effects in late fall and
winter from evaporative cooling towers.

It is of interest to consider the basis for the selection of the air-cooled condenser plants that have been
built. The well-known pioneer plant at Rugeley was
built primarily as a developmental effort-not on the
basis of economic advantage-as there was adequate
makeup water available at a reasonable price for an
evaporative cooling tower. But the plant at Ibbenbüren,
near a source of low-quality fuel, was selected on the
basis of a demonstrated economic advantage over an
evaporative cooling tower plus the cost of providing
makeup water; the unit is operated at baseload in
preference to older plants cooled by evaporative cool-

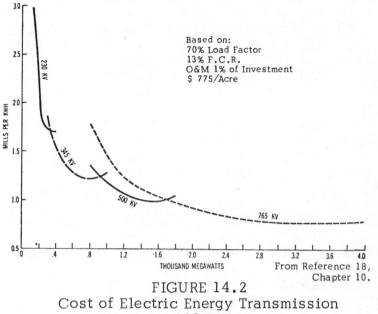

Based on:
70% Load Factor
13% F.C.R.
O&M 1% of Investment
$ 775/Acre

From Reference 18,
Chapter 10.

FIGURE 14.2
Cost of Electric Energy Transmission
For 200 Miles

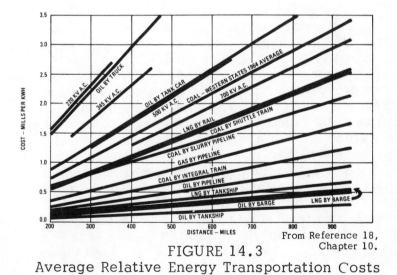

From Reference 18,
Chapter 10.

FIGURE 14.3
Average Relative Energy Transportation Costs

ing towers. The plants at Utrillas and at Grootvlei,
both in remote arid areas and near sources of low-cost
fuel, were also reportedly selected strictly on the basis
of engineering/economic comparisons of alternatives,
and so was the similarly situated, smaller plant in the
United States at the Black Hills Power & Light Company,
South Dakota. Little is known of the factors favoring
the choice of the air-cooled plants in Hungary and the
USSR, but they also were reportedly selected on similar bases.

    Situations will unquestionably arise where engineer-
ing/economic considerations will favor baseload air-
cooled condenser plants, but their widespread acceptance should be given careful scrutiny, particularly for
the Light Water Reactor nuclear plants and for reasons
of reducing lead time and interest during construction
by sidestepping regulatory procedures and objections
relating to thermal effluents. The overall increase in
waste heat rejected at the higher back pressures by
the already inefficient L.W.R. plants renders them
poor candidates for air cooling, while the current
H.T.G.R. plants with efficient steam cycles and future H.T.G.R. plants with direct helium cycles are
more amenable to air cooling, as discussed in the chapter on Nuclear Power Plants.

14.4 More Attractive Prospects for Air-Cooled Conden-
       sing System Applications

Probably a strong case can be made for site manage-
ment, as proposed in the power plant siting bills to
the Congress, with baseload units located where efficient once-through cooling-with adequate environmental safeguards-is available, or by cooling efficient baseload plants with mixed evaporative-dry
towers that will eliminate the plume and reduce relative moisture content of the discharged air. The air-
cooled plants should find widespread application
where efficiency is less important, as in intermittent
mid-range and peaking applications and in locations
near load centers. The mid-range of the load duration curve is now served by older plants, mostly at

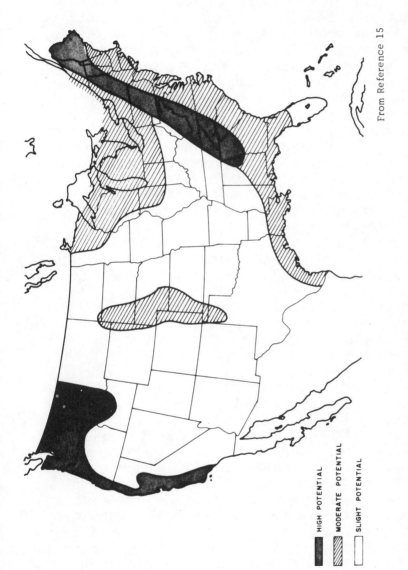

FIGURE 14.4

Geographical Distribution of Potential Adverse Effects from Cooling Towers Based on Fog, Low-Level Inversion, and Low Mixing Depth Frequency

From Reference 15

very desirable sites with ample supplies of circulating
water that could be utilized for once through cooling of
new, efficient baseload plants, by new, lower efficien
cy steam peaking/cycling units that similarly are occupying sites desirable for efficient baseload plants, and
by gas turbines often operating beyond the limited hours
that their high fuel and maintenance costs and poor efficiency warrant.

Steam peaking/cycling units equipped with air-cooled
condensers can be designed to supply the mid-range of
the load duration curve at low investment and low operating cost, as discussed in Chapter 11. Such plants, by
the nature of the load served (often the summer air-con-
ditioning demand), will have to maintain capability at
high ambients.

This is not particularly difficult to design into such
plants--with Ibbenburen maintaining full capability to
95°F, and Utrillas to 86°F--and can be provided at
much lower cost than that of the replacement gas turbine capability referred to in the studies in Table 14-II
The choice of gas turbine replacement capability at
higher ambients is surprising as the gas turbine is more
susceptible to loss of capability at high ambients than
the air-cooled steam power plant.*

Full capability of the air-cooled condenser plants at
high ambients can be provided for by shutting off feedwater heaters, by additional boiler capability for increased steam flow to the steam turbine at high ambi-
ents-the additional steam introduced after the first few
high pressure turbines stages-and by a means for providing low cost peaking capability in baseload plants.
Capability can also be maintained at high ambients by
improving the performance of the air-cooled condenser
and maintaining back pressure by spraying water into
the air stream to the heat exchangers, using a fraction
of the water that would have been required for totally
evaporative cooling of the plant.

Other applications of air-cooled condensing systems
will be for power generating water wall furnace incinerators, an area of increasing importance. The water wall

*See Section 11.3

furnace incinerators, utilizing supplemental fuel for the combustion of the 6,000 to 7,000 BTU/lb. refuse fuel and producing steam, are to be located by necessity near the cities that they will be serving, with waterfront sites or cooling water for condensing the steam unavailable.

Another area of application for air-cooled plants will be with smaller systems such as municipalities or industrial complexes that generate their own power, often in combination with district heating. The steam generated will be used for heating, air conditioning or process purposes, with the air-cooled condenser used for the process or seasonal surplus.

Further possibilities for air-cooled condensing systems may develop with the solar and geothermal power plant concepts currently under study with Federal support. Both types of plant will require means for rejecting heat, and, if practical power plant designs evolve from such efforts, it is unlikely by the nature of the plants that potential sites will have ample supplies of cooling or evaporative water available.

REFERENCES

1. Oplatka, G., Economic Expansion of a Power Supply Network; Brown Boveri Review, April 1969.

2. Melentev, L. A. and Lavrenenko, K. D., Selection of Effective Equipment for Operation to Cover the Variable Part of the Load Graph of Power Systems; Thermal Engineering (Teploenergetika) March 1971, HRVA Translations.

3. Carlin, J. F. et al, Corporate Model   of an Electric Utility; IEEE Spectrum, June 1969.

4. Frazer, R. E. and Ranson, R. C., Is Interest During Construction "Funny Money"?; Public Utilities Fortnightly, December 21, 1972.

5. Chave, C. T., Applicability of Air Cooling to Siting Problems in the Pacific Coast Area; Paper presented to the Pacific Coast Electric Association, March 27, 1970, San Francisco, California.

6. Forgo, L., Air Cooled Condensing Equipment for Thermal and Nuclear Power Stations as a Protection of the Environment; Paper 2.3-167, The Eight World Energy Conference, Bucharest 1971.

7. Lowe, W. W., Creating Power Plants: The Costs of Controlling Technology, in "Energy Technology to the Year 2000"; Technology Press, Cambridge, Mass.

8. Katz, M., Decision Making in the Production of Power, Scientific American, September 1971.

9. Jaske, R. T., A future for Once-Through Cooling; Power Engineering, January and February 1972.

10. Light on Power; N. Y. Times Editorial, May 20th, 1972. The current legislative status of proposed bills relating to power or fuels is reported in "Congressional Scoreboard", frequently appearing in "Public Utilities Fortnightly".

11. Engineering for Resolution of the Energy-Environment Dilemma; Report by the Committee on Power Plant Siting, National Academy of Engineering, Washington, D. C., 1972.

12. Considerations Affecting Steam Power Plant Site Selection; A Report sponsored by the Energy Policy Staff, Office of Science and Technology. U.S. Govt. Printing Office 1968.

13. Coal Companies Rediscover the West; Business Week, December 11, 1971

14. Aynsley, E., Cooling Tower Effects: Studies Abound; Electrical World, May 11, 1970.

15. Potential Environmental Modifications Produced by Large Evaporative Cooling Towers; E. G. & G. Inc. Report for the Water Quality Office, January 1971, Environmental Protection Agency, The Government Printing Office, Washington, D. C., 16130 DNH 01/71.

16. Are Wet/Dry Cooling Towers the Answer? Electrical World, September 15, 1972.

17. Smith, H. L., Changing Generation Patterns: Power Engineering, November 1970.

18. Study Favors Plants in Outer Urban Area; Electrical World, January 15, 1971.

19. Witt, J. A., Discussion to "Rugeley Dry Cooling Tower System" by Christopher, P. J. and Forster, V. T., Proceedings of the Institutions of Mechanical Engineers, Volume 184, Part I, No. 11, 1969-70.

20. Bender, R. J., Steam Generating Incinerators Show Gain, Power, September 1970.

21. Leung, P., Reti, G. R. and Schilling, J. R., Dry Cooling Tower Plant Thermodynamic and Economic Optimization; A.S.M.E. Paper 72-Pwr-5.

22. Drying Up the Cooling Cycle; Electric Light and Power, November 1972.

23. Energiepolitik-Trockene Ruckkuhlung in Nordrhein Westfalen (Dry Cooling in Nordrhein Westfalen); Report by the Battelle-Institute e.v., Frankfurt am Mein, issued by the Ministry, 1971.

STUDIES REFERRED TO IN TABLE 14-II.

"A." Smith, E. C., and Larinoff, M. W., Power Plant Siting, Performance and Economics with Dry Cooling Tower

Systems, Proceedings of the American Power Conference, Volume 32, 1970.

"B." Research on Dry-Type Cooling Towers for Thermal Electric Generation, Part I, R. W. Beck and Associates Report for the Water Quality Office, Environmental Protection Agency, November 1970, Government Printing Office, Washington, D.C. The conclusions of the study are also given in "Electric Power Generation with Dry-Type Cooling System" by Rossie, J. P., et. al. Proceedings of the American Power Conference, 1971, Volume 33, and are also summarized in "Dry Cooling Tower Shows Promise", Electrical World, June 1, 1971.

"C." Rossie, J. P. and Williams, W. A., The Cost of Energy from Nuclear Power Plants Equipped with Dry Cooling Systems; A.S.M.E. Paper 72-Pwr-4.

"D." Oleson, K. A., and Silvestri, G. J., Dry Cooling For Large Nuclear Power Plants; Westinghouse Power Generation Systems Report No. Gen-72-004, February 1972. Also summarized in Dry Cooling affects more than costs. Electrical World, July 1, 1972.

"E" &"F." Von Cleve, H. et al, Economics and Operating Experience with Air Cooled Condenser; Proceedings of the American Power Conference Volume 33, 1971.

"G." Ritchings, F. A. and Lotz, A. W., Economics of Closed vs Open Cooling Water Cycles; Proceedings of the American Power Conference, Volume XXV 1963; also in Power Engineering, May and June 1963.

TERMINOLOGY

General on Power Plant Cooling Systems:

Waste heat rejection from a power plant is a second law of thermodynamic necessity in converting heat into mechanical work.

In a "dry cooling system" no water is evaporated, and the heat rejected from the cycle is dissipated to the air. In a "wet cooling system" (wet cooling tower or spray pond) the circulating cooling water is sprayed into the airflow and part of it, at a rate roughly equal to the condensing steam flow, is evaporated so that not only heat but a substantial amount of moisture is also added to the atmosphere.

With dry cooling systems the condensation of steam can be:

Direct: Condensation in radially finned tubes with airflow on the outside.

Indirect: Condensation in a spray condenser; from the condenser hot well the flow is divided into feedwater to the boiler and circulating water to be cooled by air, the ratio of the two flows in the order of 1:30 to 1:60. The indirect dry cooling system is often referred to as the "Heller System."

Both dry and wet cooling systems, sometimes referred to as "closed cycle cooling systems," transfer all the power plant waste heat to the atmosphere, in contrast to "open" or "once through" cooling systems that transfer the power plant waste heat to a large body

of water such as a lake*, river, or the sea. The waste
heat is eventually dissipated by convection and evaporation to the atmosphere but for a varying fraction
(depending on survace area and water temperature,
weather conditions, cloud cover, etc.) that is lost to
space by long wave radiation.

The contrasting difference between power plants
with air-cooled condensing systems and the other two
common modes of power plant cooling is that no water
is evaporated; that is, the waste heat is rejected to
the atmosphere without consumptive water usage or
moisture addition to the atmosphere.

Perhaps a third possible mode of waste heat rejection from a power plant should be mentioned, that
by direct radiation to space. Extraterrestrial power
plants, lacking a large convective surrounding medium to absorb the discharged waste heat are designed
to radiate it to space through a high temperature condenser, or radiator, with considerable reduction in
efficiency because of the high temperature at which
the heat is rejected.

Some of the terms in power plant practice used
in the text are briefly defined below:

Air-Cooled Condenser. A finned tube heat
exchanger condensing steam from a power plant's turbine by transferring the rejected heat to the air.

Air Heater or Air Preheater. Heat transfer apparatus through which combustion air is passed and heated
by a medium at higher temperature, such as the products

---

*A small lake that receives the power plant circulat-
ing water and dissipates the waste heat by evaporation, without mixing with a large body of water, will
qualify as a closed system. Such captive lakes,
with power plant output to lake surface ratio of $\frac{1}{2}$ MW
to 3 MW/acre, have a surface temperature considerably above their natural equilibrium temperature.

of combustion, condensate or steam. In the regenerative type, or Ljungstrom, heat transfer is accomplished by a multiplated element rotating in the combustion air and flue gas streams.

Alkalinity. The amount of carbonates, bicarbonates, hydroxides and silicates or phosphates in the water, reported as grains per gallon or ppm as calcium carbonate. Also the predominance of negative hydroxyl ions over positive hydrogen ions in a solution; the logarithm of the reciprocal of the latters' concentration, in grams per liter, referred to as the pH of the solution.

Allowable Working Pressure. The maximum pressure for which the apparatus was designed and constructed to operate. It is usually tested, once or periodically, to some higher pressure.

Annulus Area. The last turbine stage exit area.

Approach. In wet cooling system practice, "approach" to the wet bulb is the difference between circulating water temperature leaving the tower and ambient wet bulb temperature. In dry cooling system practice, and usually only with direct steam condensation, "approach" to the dry bulb is the difference between the steam temperature at the turbine exhaust flange and the ambient dry bulb temperature; more often the term "initial terminal difference," or "ITD", is used in this context. "Approach" to the dry bulb with an indirect condensation system is sometimes used as above, but more often the term "tower approach" is defined, by analogy with wet cooling systems, as the temperature difference between water leaving the tower (and entering the spray condenser) and the ambient dry bulb temperature. A closer "approach" always entails an increase in heat transfer surface and cooling system investment cost.

Auxiliary Equipment (of a Generating Station). Accessory equipment necessary for the operation of a power station, such as pumps, fans, coal pulverizers, etc.

Available Energy. The isentropic enthalpy difference between a given initial steam condition and a specified final pressure.

Back Pressure. The pressure prevailing at the turbine exhaust.

Bare Tube Surface. The outside tube surface of the tubes of a finned heat exchanger, excluding the area of the fins.

Base Load. The minimum electrical load during a given period of time, usually a year.

Base Load Station. A generating station which is normally operated to take all or part of the base load of a system and which, as a result, operates essentially at constant output.

Binary Cycle. A thermodynamic cycle utilizing two working fluids. The first receives heat at high temperature and after expanding to generate power, rejects heat at some intermediate temperature to a second working fluid which expands to a lower temperature generating power and rejecting heat to the environment.

Blowdown. Removal of a portion of a boiler's or wet cooling tower's water for the purpose of reducing concentration of undesirable constituents.

Boiler Feed Pump. The pump used to raise condensate from some intermediate pressure, usually deaerator pressure, to slightly above boiler drum pressure.

Breeching. A duct for the transport of combustion gases between parts of a steam generating unit or to the stack.

Bunker C Oil. Refinery residual fuel oil, usually of considerable viscosity, used in power plants or marine boilers.

Bus Bar Cost. The cost of generating electricity, usually in mills (tenths of a cent) per kilowatt hour. It includes fuel and operating costs plus the fixed charges on the generating plant, including the main (step-up) transformer.

Capability. The maximum load which a generating unit or apparatus can carry under specified conditions for a given period of time without exceeding design limits of temperature and stress. Sometimes the term "capacity" is also used in the same context.

Capability Margin. The difference between net system capability and system maximum load requirements.

Capacity Factor. The load or output of a machine or power plant over a period of time divided by the output, had the apparatus operated at its maximum capability.

Circulating Water. The water flow to which the latent heat of the condensing exhaust steam is rejected.

Closed Cycle Cooling. Described under "General on Power Plant Cooling Systems" at the beginning of this section.

Combined Cycle. An arrangement utilizing gas turbine(s) and steam turbine(s) where the gas turbine(s) and steam generator(s) are on a common air or combustion products circuit.

Condensate. The high purity water resulting from the condensation of steam.

Condensate Pump. A pump used to pump condensate usually from subatmospheric pressure to deaerator pressure or to the inlet of the boiler feed pump.

Condensate Reheat. The use of the kinetic energy of the exhaust steam into the condenser to raise condensate temperature above the saturation temperature

corresponding to the pressure of the turbine exhaust
flange. More often this energy is used in combating
condensate subcooling, defined below. Sometimes
the term describes the use of exhaust or extraction
steam to raise condensate temperature to saturation
for deaeration in the condenser.

Condensate Subcooling. (Also referred to as
"Condensate Depression.") The lowering of the condensate temperature below the saturated steam temperature corresponding to the turbine exhaust pressure,
indicative of poor performance of the cooling system
equipment. The additional temperature difference is
not utilized in heat rejection to increase work by the
turbine; instead it results in additional heat input
required into the cycle for heating the condensate.

Condenser. A heat exchanger where exhaust
steam is condensed and heat rejected to another fluid
either in tubes (surface condenser) or by mixing
(direct contact or spray condenser).

Condenser Terminal Difference. A term used with
surface condensers to indicate the difference between
circulating water leaving the condenser and saturated
steam temperature (which is also the condensate temperature in the absence of subcooling); 5°F is the
practical minimum difference specified. Terminal
Difference is inevitable with the crossflow arrangement between steam and circulating water employed
in surface condensers, but such is not the case with
spray condensers, where circulating water and condensate mix with terminal difference and subcooling
being identical; as a result terminal difference should
be held close to zero. The HEI "Standards for Direct
Contact Barometric and Low Level Condensers" (Fifth
Edition, 1970) recommends a rather generous 3°F minimum Terminal Difference.

Cooling Range. The temperature difference to
which water is cooled in a wet cooling system or in an
indirect condensation dry cooling system tower.

Crude Oil. Unrefined petroleum.

Cycling Unit. A power generating unit used to cover peak/intermediate duration demand and as a result shut down and restarted frequently, sometimes daily, in contrast to the continuous operating base-load units.

Deaeration. Removal of dissolved air and gases from condensate and boiler feedwater by heating to saturating temperature at the corresponding pressures.

Deaerator. A direct contact feedwater heater, where steam extracted from the turbine heats conden-sate to saturation temperature, releasing air and en-trained gases prior to its introduction into the boiler. Deaeration also takes place in the condenser.

Demand. The rate at which electric energy is required by the power consuming equipment of its customers; the load.

Distillate Fuels. Liquid fuels distilled from crude petroleum, with the heavier No. 5 and No. 6 (residuals) remaining behind.

Dry Bulb. The ambient air temperature.

Dry Cooling System. Described under "General on Power Plant Cooling Systems" at the beginning of this section.

Dry Cooling Tower or Dry Tower. A term some-times uded to describe a mechanical or natural draft tower for an indirect dry cooling system.

Economizer. Heat recovery equipment designed to transfer heat from the products of combustion to boiler feedwater.

Electrostatic Precipitator. Equipment used for collecting dust from combustion products by charging the dust particles as they travel in an electric field and collecting them on the positive electrodes (plates).

Excess Air. Air supplied for combustion in excess of the theoretical air required.

Exhaust Loss. Losses in the energy of the steam between the last turbine stage and the condenser, the major part due to the kinetic energy of the steam.

Extended Surface. Finned heat exchanger surface.

Expansion Line. The line on the Mollier diagram (enthalpy vs. entropy) through the points specifying the condition (status) of steam as it expands through the turbine.

Extraction. The removal of steam from the turbine before it has fully expanded to condenser pressure for feedwater heating or other purposes.

Feedwater. Water introduced into the boiler, made up of condensate and makeup; also a term used to describe condensate after the boiler feedpump.

Feedwater Heater. Heat exchanger apparatus, of the shell and tube or the direct contact type, used to heat up condensate (low pressure heaters) or feed-water (high pressure heaters).

Fixed Charge . The annual costs arising from the ownership of property: depreciation, taxes, insurance and the cost of money (interest) used in purchasing the property.

Flue Gas. The combustion gases after they leave the steam generator on the way to the stack.

Heat Rate. The measure of a generating station's or a turbine's efficiency expressed in BTU's per kilo-watt hour, as heat input divided by electrical output, usually specified as net or gross; the inverse expressed in percent (1 kWhr = 3412.7 BTU) will be the respective efficiency.

Heller System. The indirect dry cooling condensation system employing a spray condenser pioneered by Professor Heller of Hungary.

Indirect Condensation System. Described under "General on Power Plant Cooling Systems" at the beginning of this section.

Jet Condenser. Synonymous with "spray condenser."

Fuel Cost. The cost of fuel delivered to a utility or a power station, in $/ton (coal), $/bbl (oil), or ¢/MCF (gas); fuel cost is often expressed in ¢/MMBTU of heat content. The term is also used to express the fuel cost component (in mils per kWhr) of the busbar cost of electricity; in this case the efficiency of the generating unit is also used in calculating the fuel cost.

Generator Cooling. The removal of the heat generated by electrical losses in the generator. In ratings up to 50 MW this is usually accomplished by direct forced air cooling; pressurized hydrogen consequently cooled by water is used for higher ratings, while for ratings approaching 1000 MW and above, direct liquid cooling of the generator is used.

High Pressure Section. The turbine section first admitting the steam from the boiler.

Higher Heating Value. The heat given out during combustion of a fuel, including the latent heat of any water vapor present; HHV is used in power plant practice while the LHV (which excludes the latent heat of water vapor) is often used with gas turbines and internal combustion engines.

Hood. The passage between the last turbine section and the condenser.

Initial Terminal Difference (ITD). The difference between condensing steam temperature corresponding to turbine exhaust pressure and the ambient temperature.

Intermediate Pressure Section. The turbine section admitting the steam after it has expanded through the high pressure section; it is followed by the low pressure section(s).

Interest During Construction. The accumulating cost of money while a generating or other facility is in the process of construction. Upon commissioning, the cost of money for the facility, along with the other fixed charges, enters the rate base (IDC capitalized). In another accounting treatment IDC is annually charged off as expense.

Load. The amount of electric power required by the customers at a given time, also referred to as Demand.

Load Curve. The curve showing the load vs. time for the specified period (day, week, month or year).

Low Pressure Section. The turbine section receiving the low pressure steam; from the L.P. Section the steam discharges to the condenser.

Main Stop Valve. The isolating valve between steam generator outlet and turbine inlet.

Makeup. The water added to a boiler, or wet cooling tower circuit, to replace water lost through blowdown and leakage, or blowdown, evaporation, leakage and drift in the latter case.

Mechanical Draft. The generation of airflow by fans.

Moisture Region. The Mollier chart region where the vapor and liquid states coexist (under the saturation line).

Multipressure Condenser. The arrangement where circulating water--or mixed circulating water condensate in the case of spray condensers-- travels successively through two or more condenser regions

(or separate shells) condensing steam at successively
higher pressures. The arrangement offers a thermodynamic
advantage in reducing irreversibility in the heat rejection process.

Natural Draft. The generation of airflow by ex-
ploiting the difference in air density due to differences
in air temperature when discharging heat to air (the
so-called chimney effect). In wet natural draft towers
the increased moisture also contributes to the draft.

Once-through Cooling. Described under
"General on Power Plant Cooling Systems" at the beginning of this section.

Open Cycle Cooling. Used interchangeable
with "once-through cooling."

Peak Load. The maximum load to be satisfied
by a system, usually occurring for a given short period
of time.

Prime Mover. The engine or turbine that drives
an electric generator.

Pumped Hydro. Electric generating capability
by hydraulic turbines usually for peak load periods, the
reservoir being recharged by reversing the turbines and
pumping back the water during periods of low demand.
Also referred to as "Pumped Storage."

Rate Base. The value established by a regulatory
agency for a utility's property (generating plants, transmission lines, distribution networks, etc.), upon
which a specified rate of return is allowed. Other items
such as depreciation methods, treatment of taxes, etc.,
are also often regulated in conjunction with the rate
base.

Regenerative Feedwater Heating. The practice
of using partially expanded steam from the turbine
(extraction steam) to heat up condensate in feedwater
heaters, before the boiler, raising the average heat

addition temperature to the cycle and improving efficiency; other benefits also accrue.

Reheat. The practice of reheating the steam between the high pressure and intermediate pressure turbine sections by taking it back to the steam generator's "Reheater." Reheat improves efficiency by increasing the amount of heat added at the higher temperature; it also reduces undesirable moisture in the last stages of the steam's expansion to the condenser pressure. (Two reheat stages are sometimes used.)

Reserve Margin. The difference between net system capability and maximum (peak) system demand, a margin intended to meet unanticipated demand and equipment breakdowns.

Top Heater. The last (and highest pressure) feedwater heater before the feedwater enters the steam generator's economizer.

Spinning Reserve. Generating units (and their capability) carrying little or no load but synchronized and ready to take up load in the event of unanticipated demand or malfunction of operating units; or a unit operating for the same purpose below its maximum capability.

Spray Condenser. A heat exchanger where turbine exhaust steam condenses by direct contact with sprayed circulating water. In current practice synonymous with "Direct Contact Condenser" or "Jet Condenser," although in past practice the terms were not synonymous.

Summer Peak. Maximum demand occurring in the summer. More frequent in recent years as air conditioning loads have increased.

Surface Condenser. A condenser utilizing tubes. The condensing steam (condensate) and the cooling medium (circulating water) are separated by the tube surface and do not mix in this case. Also as there is no mixing, a temperature difference exists between the steam and circulating water leaving the condenser.

System Load. The combined electrical demand by a utility or system's customers.

Terminal Difference. Described under "Condenser Terminal Difference."

Tower. The term is used indiscriminately to denote a wet or dry cooling system installation of the mechanical or natural draft type.

Tower Approach. Described under "Approach."

Turbine Exhaust Flange. The point where the exhaust turbine casing ends and the condenser shell is connected; also referred to as "condenser flange." Used as a reference point in designating exhaust steam properties and turbine and condenser performance.

Waste Heat. The substantial latent heat of condensation of steam rejected to the environment but which in certain cases, such as use of the turbine exhaust steam for heating buildings or process applications, can be beneficially utilized. The term is also applied to the above ambient heat rejected with the combustion gases.

Winter Peak. Maximum demand occurring in a winter month, usually in the early evening in late December because of the heavy lighting load at such time.

Wet Bulb. The temperature to which ambient air can be adiabatically cooled to saturation by the addition of water vapor, usually measured by a thermometer the bulb of which is kept wet by a moist wick.

Wet Cooling Tower. A mechanical or natural draft cooling tower in a wet cooling system where the circulating water is cooled by the evaporation of a fraction of it in the airflow.

The following sources were found useful in compiling the above list of terms:

GLOSSARY OF ELECTRIC UTILITY TERMS, prepared
by the Statistical Committee of the Edison Electric
Institute, 750 Third Avenue, New York, N.Y.

GLOSSARY OF TURBINE TERMS, in "Steam Turbine
Performance and Economics", by R. L. Bartlett, Mc-
Graw-Hill Book Co., Inc., New York, 1958.

GLOSSARY OF BOILER AND ELECTRIC UTILITY TERMS
(from Lexicon of American Boiler Manufacturers Association) in "Combustion Engineering", Revised Edition
G.R. Fryling, Editor. Published by Combustion
Engineering, Inc., New York, N.Y.

GLOSSARY OF ABBREVIATIONS AND DEFINITIONS,
The 1970 National Power Survey; Part I. The Federal
Power Commission, U.S. Government Printing Office,
Washington, D.C.

Appendix A

AIR-COOLED POWER PLANT STATISTICS

Chronological information and information on performance
and hardware design values for several air-cooled condensing system installations is contained in Table A-1, A-II
and A-III from the references given in several chapters
and other sources; the values for the mechanical draft
systems have been obtained from proposals submitted by
manufacturers. The "year" given for an installation applies
to completion or first commercial operation, but there appears
to be some discrepancy between various sources as well
among other data.

In Figure A.1 air-cooled power plant size growth is presented
pictorially by plotting largest plant size vs. year of operation
from Tables A-1 and A-2. A similar trend is shown in Figure
A-2 that presents total installed air-cooled plant capacity
and includes the plants from Tables A-1 and A-2 and the
many smaller direct steam condensation air-cooled power
plants.

Air-Cooled Power Plant Statistics

TABLE A-I

DIRECT AIR-COOLED STEAM CONDENSING INSTALLATIONS

Year	Location	Condensing Capacity (& Power Output)	Design Air Temperature	Condensing Pressure
1939	*Ruhr, Germany. Coal Mine Power Station	12,100 lbs/hr (MW)	54 F	1.73 " Hg
1944	* Mobile, Locomotive-mounted Power Trains Developed in U.S.	(5 MW)		
1956	Dudelange Steel Works in Luxembourg	110,000 lbs/hr (13 MW)	54 F	2.25" Hg
1956/57	*Rome, Italy. Municipal Power Station	2 x 150,000/hr 190,000 lbs/hr (2 x 29/36 MW)	57 F	1.80" Hg
1962	Black Hills Power & Light, Sough Dakota. Conversion	(3 MW)		
1960/61/65	* Volkswagen Works Wolfsburg, Germany	3 x 242,000/ 286,000 lbs/hr (3 x 40/48 MW)	F	2.7" Hg
1968	Bremen, Germany. Municipal Incinerator	220,000 lbs/hr	81 F	212 psia
1969	Black Hills Power & Light, Sough Dakota	153,000 lbs/hr (20 MW)	80 F	4.7" Hg
1969	Utrillas, Spain. Mine Mouth Power Station	692,000/768,000 (146/160 MW)	59 F	2.9" Hg
1977	* Pacific Power & Light & Black Hills Power of Light Dakota Wyoming Coal Fields	(330 MW)		up to 15" Hg
1977	* Germany. Nuclear	(300 MW)		

* Asterisk indicates plant is included in Figure A.1

TABLE A-II

INDIRECT AIR-COOLED STEAM CONDENSING INSTALLATIONS

Year	Location	Condensing Capacity (& Power Output)	Design Air Temperature	Condensing Pressure
1955	*Hungary. Prototype	11,000 lbs/hr (1.2 MW)		
1961	Danube Steel Works. Hungary	105,000 lbs/hr 13/16 MW)		1.2" Hg
1961	* England. Rugeley	(120 MW)	52 F	1.2" Hg
1964	Quetta Power Station Pakistan	(2 x 7.5 MW)		
1967	* Ibbenburen Germany	660,000 lbs/hr	35 F	1.3" Hg
1969	Gyongyos Hungary	(2 x 100 MW)		
1971	Grootvlei South Africa	(200 MW)		6.6" Hg
1970/71/73	* Razdan, U.S.S.R.	(3 x 220 MW)		
1976	* Soviet Armenia Nuclear	(2 x 400 MW)		

*Asterisk indicates plant is included in Figure A.I

TABLE A-III
DESIGN DATA ON INDIRECT
AIR-COOLED CONDENSING SYSTEMS

	NATURAL DRAFT	
	RUGELEY	IBBENBUEREN
HEAT REJECTED	587 10^6	650 10^6
ITD, °F	33 *	51 *
TOWER SIZE OR AREA REQUIRED	350' x 325' DIA.	327 x 265 DIA.
NO. AND TYPE OF EXCHANGER ASSEMBLIES	217 (3 x 16') x 8' x 6" 80,000 ft^2 FRONTAL AREA	166
TUBE LENGTH	3 x 16'	3 x 16'
TUBE DIAMETER	.59" I.D.	.59
TUBE METAL	ALUMINUM	ALUMINUM
FINN TYPE & METAL	FORGO-HELLER ALUMINUM PERFORATED PLATE FIN	FORGO-HELLER ALUMINUM PERFORATED PLATE FIN
WATER PASSES	2	2
TOWER CIRC. WATER, GPM	73,250	66,500
WATER TO STEAM RATIO	64:1	50.4:1*
WATER RISE, °F	16	19.2
WATER VELOCITY, FPS	3.8	3.9*
AIR FLOW, CFS	384,000*	318,000
AIR VELOCITY, FPS	4.8 NORMAL TO COOLERS	5.0*NORMAL TO COOLERS
AIR RISE, °F	23	29.6*
AIR SIDE PRESSURE DROP, INCHES W.G.	ABOUT .2"	
BARE TUBE, FT^2	378,000 (INSIDE)	296,000 (INSIDE)
U BASED ON BARE SURFACE	117	114
EFFECTIVE MTD IN CROSSFLOW, F	13.2	19.2
RATIO OF FINNED TO BARE SURFACE		
U BASED ON TOTAL SURFACE		

* Asterisk indicates that the figure given has been calculated or inferred from other data.

TABLE A-III
DESIGN DATA ON INDIRECT
AIR-COOLED CONDENSING SYSTEMS

(continued)	MECHANICAL DRAFT	
DESIGN "A"	DESIGN "B"	DESIGN "C"
1000 10^6	3800 10^6	1000 10^6
35	62	48.8
	16 ACRES	
	624	136
	36" x 6'-9" x 9"	40' x 8'-10" x 9"
36'	36'	40'
1" O.D., 16 BWG	1" O.D., 16 BWG	1" O.D., 18 BWG
STEEL	STEEL	ADMIRALTY
PLATE ALUMINUM 9/INCH	AL 2-1/4 O.D.	AL 2-1/4 O.D.
	10/INCH	11/INCH
2	2	2
111,100	422,000	133,700
55:1	52.7:1	
18	18	15
1.9	5.2	5.9
750,000	1,900,000	584,000
6.7 FACE VELOCITY	12.6	12.2
20.6	36	27.1
.29"	.37	.25
589,000	846,000	268,000
(OUTSIDE)	(OUTSIDE)	(OUTSIDE)
111	144	165
15.4	31.1	22.8
21.6	18.7	22
4.9	7.4	7.1

Air-Cooled Power Plant Statistics

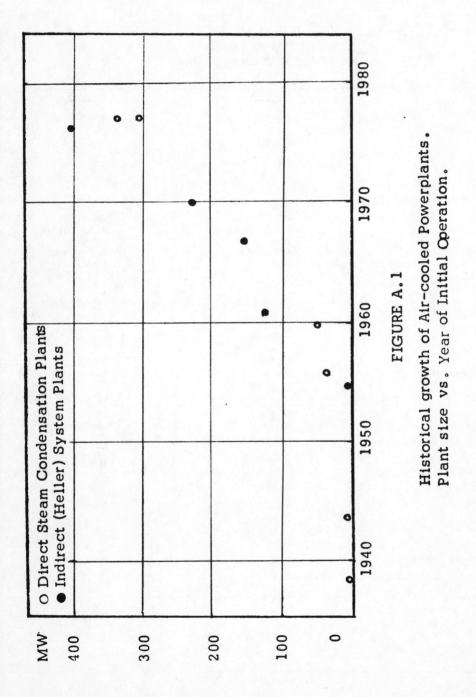

FIGURE A.1

Historical growth of Air-cooled Powerplants.
Plant size vs. Year of Initial Operation.

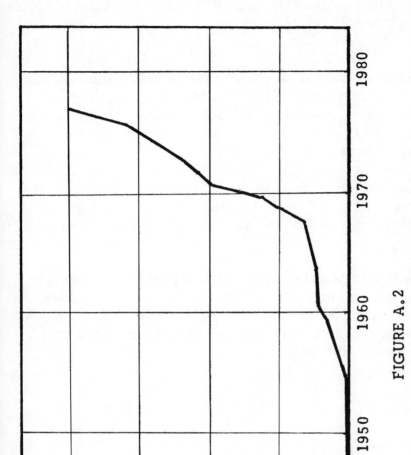

FIGURE A.2

Total Installed Air-cooled Plant Capacity

Appendix B

LOSS OF PLANT CAPABILITY AND INCREASED HEAT REJECTION AT HIGHER TURBINE BACKPRESSURES

The loss of plant capability, heat rate deterioration, and increased heat rejection at the higher turbine backpressures
that will be encountered with air-cooled power plants is a
function of initial steam conditions, cycle configuration,
and turbine exhaust end selection.

In this appendix, only the effects of increased backpressure
on plant capability and heat rejection will be considered;
heat rejection as affected by other factors is discussed in
Appendix C.

For a simplified analysis of the effect of inlet and exhaust
temperatures on plant cycle efficiency, it is usual to refer
to the ideal Carnot efficiency attainable with these (absolute)
temperatures, multiplied by a constant to account for cycle
deviations and mechanical losses:

$$n = k \cdot \frac{T'_1 - T'_2}{T'_1} \qquad (B.1)$$

If T_1 and T_2 are the absolute source and sink temperatures,
and T'_1, T'_2 the cycle working fluid inlet and discharge
absolute temperatures, $T_2 - T'_2$ will correspond to the
Initial Terminal Difference (ITD), the temperature difference
between cycle fluid discharge temperature and sink
utilized across the heat exchanger surface to promote heat
transfer. Increasing the difference (raising T'_2), reduces
plant efficiency but also reduces surface, size and cost
of the heat rejection system.

Loss of Plant Capability and Increased Heat Rejection at
Higher Turbine Backpressures

The above simplified analysis results in a linear relationship between increasing plant heat rejection temperature,
reduction in efficiency and loss of capability--about .17%
loss of capability per degree F heat rejection temperature
increase for a current design, once through cooled, fossil
fired steam power plant. The analysis neglects the work
that goes into fluid volume expansion (differing for steam
to other fluids) and the kinetic energy (exhaust loss) of
the departing fluid that depends on turbine exhaust end size.

A more accurate indication of the loss of plant capability
with increasing heat rejection temperature is given by the
loss in availability of the steam, $\left(\dfrac{\Delta h}{\Delta p}\right)_s$, which includes the
work that goes into volume expansion. Loss of plant
capability and increased heat rejection on the basis of lost
steam availability is shown by curve I , Figure B.1,
for the conditions specified.

The influence of turbine exhaust end selection (e.e., the
turbine exhaust area effect on the loss of the kinetic energy
of the discharging steam) is shown by curves II and II' and
III and III' in the same figure, corresponding to the curves in
Figure 9.8b labeled 'non-condensing type"(small area,
heavily loaded exhaust end) and "condensing type" large
area, lightly loaded exhaust end). It can be seen that by
selecting the smaller, heavily loaded exhaust end and com-
promising efficiency at the lower backpressures (or ambients),
efficiency, capability and heat rejection are little affected
by rising backpressures.

Loss of Plant Capability and Increased Heat Rejection at
Higher Turbine Backpressures

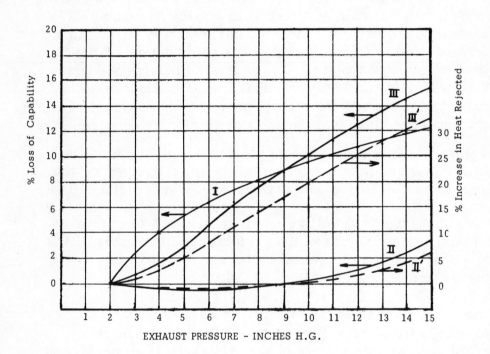

EXHAUST PRESSURE - INCHES H.G.

Curve I   Turbine Heat Rate at 2" Hg taken as 7800 BTU/KWHR
Curve II   Turbine Heat Rate at 2" Hg taken as 7800 BTU/KWHR

Curve III   Turbine Heat Rate at 2" Hg taken as 8424 BTU/KWHR

Curve III   Heat Rejected per KW at 2" Hg: 4387 BTU
Curve II   Heat Rejected per KW at 2" Hg: 5011 BTU

Figure B.1

Effect of Exhaust End on Loss of Capability and Heat Rejected

Appendix C

REDUCTION IN HEAT REJECTED WITH IMPROVED EFFICIENCY

Power plant efficiency and heat rejection is not only a function of turbine backpressure (ambient temperature and ITD), discussed in Appendix B, but also of cycle configuration, i.e., initial steam conditions, number of feedwater heaters, number of reheat stages, etc.

In Figure C.1, heat rejected, and percent change in heat rejected divided by percent change in heat rate have been plotted vs. heat rate and efficiency.

In the above figure, if the "heat rate" is taken to indicate plant heat rate, the heat rejected and its percent change will closely correspond to that rejected through the power plant cooling system for the case of a nuclear plant.

For a fossil power plant, where a substantial amount of heat is lost through the stack (boiler losses), the heat rejected through the cooling system and its change can be surmised from Figure C.1, if "heat rate" is interpreted as turbine heat rate. Typical heat rates of fossil fired and nuclear power plants are given in Table 12-I.

Reduction in Heat Rejected with Improved Efficiency

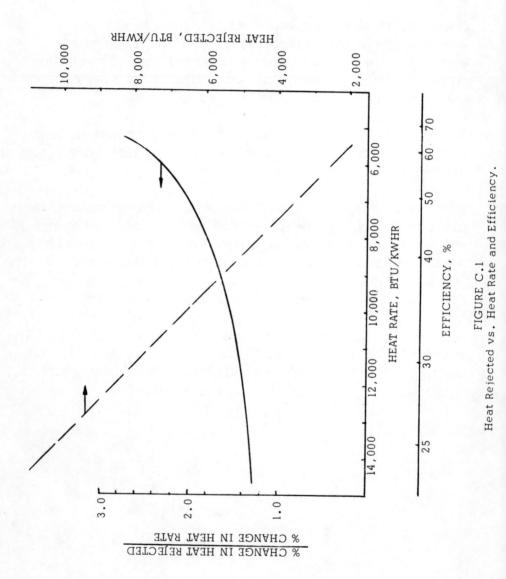

FIGURE C.1
Heat Rejected vs. Heat Rate and Efficiency.

Appendix D

SERIES (MULTIPRESSURE) ARRANGEMENT OF SPRAY AND SUR-
FACE CONDENSERS

The benefits of series arrangement for spray and surface condensers referred to in Sections 7.4. and 7.5 are further
discussed below and the equation for the reduction in condensing temperature with series spray condensers is derived.

Assuming the total heat rejected by the turbine, Q (BTU/hr),
is equally divided between in "n" spray condenser stages,
Figure D.1, the circulating water flow, w (lbs/hr), is large
in comparison to the total steam flow - a good assumption
as the circulating water to condensing steam ratio is usually
between 20 and 50 to 1, t_s the single spray condenser
condensing temperature, (°F), and t_{s2}, ... t_{sn} the condensing temperatures of the multistage condenser, Figure D.2,
the reduction in condensing temperature with the later
arrangement is derived as follows:

$$\Delta t s = t s - \frac{t s 1 + t s 2 \ldots t s n}{n}$$

$$= (t_1 + Tr) - [\ (t_1 + \frac{Tr}{n}\ ) + (t_1 + \frac{2Tr}{n}) + (t_1 + \frac{nTr)]}{n}$$

$$\Delta t_s = T_r \cdot \frac{n-1}{2n} \qquad (7.2)$$

Where T_r is the rise , (°F), in the circulating water.

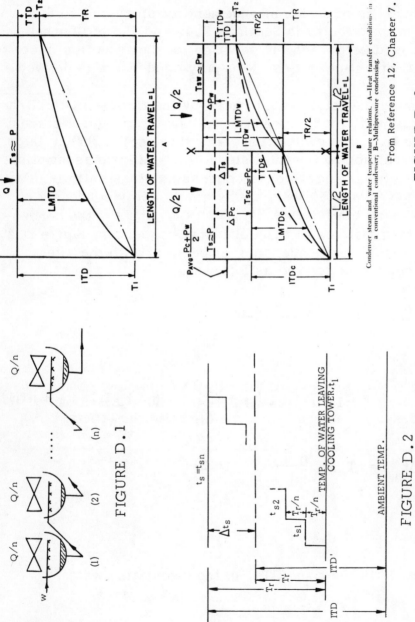

Condenser steam and water temperature relations. A—Heat transfer conditions in a conventional condenser; B—Multipressure condensing.

From Reference 12, Chapter 7.

FIGURE D.3

FIGURE D.1

FIGURE D.2

It is assumed that both the single spray condenser and the
multipressure arrangement are receiving cooled circulating
water from air-cooled heat exchanger systems of the same
size and at the same circulating water temperatures.
Circulating water temperature rise in the single condenser
and the multipressure arrangement will be the same because
of the equal heat rejected and the equal circulating water
flows; subcooling is assumed to be insignificant in both
cases.

For a two stage (two pressure) Spray Condenser the
improvement in condensing temperature will be $T_r/4$ the circulating water rise in the condenser (which is equal to the
cooling range of the tower). It can be easily shown that,
under the assumptions stated, equal distribution of the heat
rejection load between the two stages will result in the
maximum reduction in the average condensing temperature.

The reduction in the average condensing temperature with
two, three or more stages is shown in Figure 7.6. The
improvement in heat rate and capability is actually proportional not the reduction in the average condensing temperature but the reduction in the average of the corresponding
condensing pressures.

The analysis of multipressure surface condenser is far more
complicated because of the effect of the more complex heat
transfer through the tube wall – instead of by mixing as in
the spray condenser – and the effect of the terminal temperature difference, also absent from the spray condenser. For
a two pressure surface condenser, again with equally
divided heat rejection load, Figure D.3, the reduction in
condensing temperature is given by the relation:

$$\Delta t_s = \frac{1}{2}\left[(T_d + \frac{T_r}{2}) - \sqrt{T_d^2 + T_r \cdot T_d} \right] \qquad (7.1)$$

Series (Multipressure)   Arrangement of Spray and
Surface Condensers

and for a multipressure surface condenser arrangement
of "n" stages by the relation:

$$\Delta t_s = T_r \left[\frac{1}{2} + \frac{T_d}{T_r} - \frac{1}{n} \left(\frac{1}{2} + \frac{1}{\sqrt[n]{1 + \frac{T_d}{T_r}} - 1} \right) \right]$$

By employing equations 7.1 and 7.2 the improvements due
to the use of a spray condenser over a surface condenser
and their corresponding two pressure equivalents, all
serving an indirect condensation air-cooled powerplant,
will be compared.

It is assumed that in all cases actual cooling water
temperature rise – and air-cooled tower cooling range, will
be 20°F, the terminal temperature difference in the single
pressure surface condenser will be the recommended
minimum of 5°F, subcooling will be absent in the spray
condensers, and air-cooled "tower approach" (the
temperature difference between the cold water leaving the
tower and the ambient air), Figure 4.7, will be 10°F.

The required heat exchanger surface when employing the
various condenser arrangements in Table D-I has been
calculated by assuming the air-cooled system surface
inversely proportional to the initial terminal difference
(ITD) between the ambient air and condensing steam
temperature. As can be seen from the above comparison the savings in the heat exchanger surface costs
are considerable when employing multipressure spray
condensers.

As discussed in Sections 7.4 and 7.5 the above
analysis neglects the considerable effect of exhaust
loss variation and the improvement in the multipressure
surface condenser due to the reheating of the condensate
from the lower pressure stages to the condensing
temperature of the last stage.

Series   (Multipressure)   Arrangement of Spray and
Surface Condensers

TABLE D-1

Comparison of single pressure and two-pressure surface
and spray condensers.

SINGLE PRESSURE	SURFACE CONDENSERS	SPRAY CONDENSERS
ITD	35F	30F
Required Air-cooled System Surface for Constant Plant Efficiency and Output	100%	86%
TWO PRESSURE		
Reduction in Effective Cooling Range	1.9F (eq. 7.1)	5 F (eq. 7.2)
Effective ITD	33.1 F	25 F
Required Air-cooled System Surface for Constant Plant Efficiency and Output.	95%	71%

INDEX

INDEX

INDEX

INDEX